T4-AKS-183

Für Dolfi
(„Koala – Koala")
zur
Erinnerung
an
unsere
Australienreise
November 1989

von

Walter   Uta

KOALA

KOALA

AUSTRALIA'S ENDEARING MARSUPIAL

Text by

MICHAEL ARCHER, STEVEN CORK, SUZANNE HAND,
STEPHEN PHILLIPS AND MALCOLM SMITH

Edited by

LEONARD CRONIN

FOREWORD

The koala is an international envoy of the animal kingdom, warming the hearts of zoo visitors in countries as far apart as Japan and the United States. So loved is this Australian marsupial that large sums of money are spent providing the correct food and conditions for an animal that is perfectly adapted for a unique yet threatened Australian environment

Australia's own living "teddy bear", the koala is just one of our many strange and fascinating animals. Where else in the world can you find such a bizarre creature as the duck-billed platypus sharing the water with a thirsty, beautifully streamlined kangaroo, while a laughing kookaburra and a sleepy koala watch from a nearby gum tree?

Yet most people remain ignorant of the biology, behaviour, ecology and evolution of Australia's greatest tourist attraction. This is hardly surprising as this knowledge has inadvertently been a well-kept secret, confined to scientific journals and research laboratories, and the layperson has been left with fanciful stories about drunken koalas living in their gum tree havens, or sensational newspaper accounts of venereal diseases spreading through the koala populations.

It is to dispel the myths and present our knowledge of the koala to the general reader that this book has been produced. By increasing our understanding of this beautiful and endearing creature it is hoped that governments and developers will appreciate the overwhelming need to protect their habitat and conserve our rapidly diminishing native forests, some of the greatest store-houses of biological diversity left on this planet.

The first time I saw a koala in the wild it took me quite a few minutes peering into the foliage of a fairly sparse gum tree, which I had been told contained a koala, before I could actually make out what appeared to be a ball of fur propped in a fork of the branches. I remember scrambling up a nearby rock to get a closer view, and was rewarded by the opening of a sleepy eye and the laborious stretching of a hairy arm. That was it. I went back the following day, and the same koala was in exactly the same position, apparently continuing its long sleep. The next day, however, it was gone, proving that this creature really has some locomotive abilities.

The koala, it seems, is one of nature's great experiments. Life really is meant to be easy, at least for this particular mammal. With virtually no predators and hardly any competition for its food supply, the koala generally leads a sedentary lifestyle quietly munching gum leaves.

Of course, things are not quite that simple in the biological world. Eucalypt leaves are particularly tough, dry and unpalatable, which is sufficient to deter most animals, but if that is not enough, they are poisonous and barely contain enough nourishment to keep body and soul together, let alone grow and reproduce. So the koala had to come up with some remarkable adaptations to allow it to enjoy the quiet life offered by the ubiquitous gum trees.

The best way to exist on food so low in nutrition is to conserve as much energy as possible, which explains why koalas never seem to do much. As for the poisonous compounds, within that expansive midriff the koala has the most remarkable digestive system in the animal kingdom, and a liver that can deal with chemicals that would kill us in hours.

Marsupials are well known for the way they raise their young, generally keeping them tucked up in a pouch from a remarkably early age. Kangaroos can be seen bounding through the bush with a little joey sitting upright in the pouch, diving down for a drink when it gets thirsty. Wombats, on the other hand, have a pouch that faces to the rear, which makes a lot of sense for an animal that moves around on all four limbs and doesn't want its pouch scooping up all the twigs and other objects it happens to pass over. The koala, however, in its leisurely ascent into the arboreal world retains the same backward-facing pouch as its ground-dwelling relatives,

and consequently when the mother stands upright the baby koala finds itself peering down at the ground from the precarious heights of a gum tree.

In one particular zoo they even had to clip the growing infant into the pouch to stop it falling out! Fortunately this doesn't seem to happen in the wild, otherwise we would find the poor koala cubs falling from the trees like ripe fruit.

The advantages and disadvantages of life in a gum tree are clearly explained in this excellent book, written by some of Australia's leading authorities on the koala. Many newly discovered facets of the koala's biology, evolution and lifestyle are revealed, together with a new look at the evolutionary relationships within the koala family. Much more research still needs to be done to unravel all the mysteries surrounding this remarkable marsupial, for it is only through knowledge that we can ensure the future status of our unique fauna.

LEONARD CRONIN
SYDNEY, 1987

First published 1987 by
REED BOOKS PTY LTD
2 Aquatic Drive, Frenchs Forest, NSW
2086

Produced for the Publisher by
LEONARD CRONIN

© Leonard Cronin/Reed Books 1987

All rights reserved. No part of this
publication may be reproduced, stored
in a retrieval system, or transmitted in
any form or by any means, electronic,
mechanical, photocopying, recording or
otherwise, without the prior written
permission of the publishers.

National Library of Australia
Cataloguing-in-Publication data

Koala, Australia's endearing marsupial.
Bibliography.
Includes index.
ISBN 0 7301 0158 4
1. Koalas. I. Cronin, Leonard.
599.2

Designed by Warren Penney

Typeset by Times Graphics, Singapore
Printed in Hong Kong
through Imago Productions (FE) Pte Ltd

CONTENTS

INTRODUCTION TO THE MARSUPIALS

When Captain James Cook's *Endeavour* was beached at Endeavour River in 1770 in what is now Northern Queensland, his naturalists Joseph Banks and Daniel Solander collected a strange animal they believed to be unlike any other known. This was Captain Cook's Kangaroo (there is still controversy over which species it actually was) and it is often mistakenly thought to be the first marsupial discovered. In fact, Cook's expedition had collected a rabbit-like animal (a bandicoot or perhaps a kangaroo rat) and a native cat at Botany Bay some three months earlier, but they did not realise that these animals were all related.

Australian marsupials had been reported many times before this, and the first report of a marsupial from South America was made as early as 1500. Strangely, however, most early reports of marsupials were not only received with much scepticism but often did not even find their way into circulation among scientists. Thus, it was not surprising that Banks and Solander thought they had a totally new animal on their hands.

In 1500, Vincente Pinzon returned to Spain with a female opossum captured in Brazil. This he presented to King Ferdinand and Queen Isabella at Granada, drawing their attention to the remarkable pouch on its belly, into which, it is said, the royal hands were inserted. At the time the scientific world knew of no mammals of this type, and there was a narrowness of outlook which sought to categorise all new discoveries as a version of the species already known. Thus the opossum was interpreted by some as part monkey and part fox.

Pinzon's opossum was a female and had young in its pouch when captured. But the young were lost and the opossum was dead upon arrival in Spain. The world had to rely on Pinzon's description, and it seems that he believed the young were merely carried in the pouch and that the teats from which they suckled were elsewhere on the mother's body. This was the first of many misconceptions about marsupials.

There must have been other reports of these strange creatures from both Spanish and Portuguese navigators who operated in parts of the world populated by marsupials, but few survive to this day and even fewer appear to have reached the scientists of the time. One very detailed description of a cuscus, a member of the possum family, has been uncovered in the writings of a Portuguese official named Antonio Galvao who was stationed in the Moluccas in the mid 16th century.

It is in the early 17th century, however, that the first reports of marsupials in the vicinity of Australia appear. The expedition of the Spaniard Torres reported seeing what was probably a cuscus in Papua in 1606. Torres' expedition made a brief excursion into Australian waters off the coast of northern Queensland but sighted no marsupials, only many "troublesome" flies which, in the words of a member of the expedition, "seemed as if they wanted to eat the men up ...".

Between 1618 and 1627 Australia was visited by various Dutch navigators, but few natural history observations have been passed down from these. In 1628 Francois Pelsaert's ship *Batavia* was wrecked at Houtman's Abrolhos, off the coast of Western Australia, and he gave the first detailed description of an Australian marsupial. The animal Pelsaert described was in fact the tammar wallaby *Macropus eugenii*. His original description indicates how strange he found these beasts to be:

"Besides we found in these islands large numbers of a species of cats, which are very strange creatures; they are about the size of a hare, their head resembling the head of a civet-cat; the forepaws are very short, about the length of a finger, on which the animal has five small nails or fingers, resembling those of a monkey's forepaw. Its two hind legs, on the contrary, are upwards of half an ell in length and it walks on these only, on the flat of the heavy part of the leg, so that it does not run fast. Its tail is very long, like that of a long-tailed monkey; when it eats, it sits on its hind legs, and clutches its food with its forepaws, just like a squirrel or monkey."

But it was their manner of reproduction which Pelsaert found most amazing, and his descriptions and interpretations illustrate another odd misconception about marsupials.

"Below the belly the female carries a pouch, into which you may put your hand; inside this pouch are her nipples, and we have found that the young ones grow up in this pouch with the nipples in their mouths. We have seen young ones lying there, which were only the size of a bean, though at the same time perfectly proportioned, so that it seems certain that they grow there out of the nipples of the mammae..."

Whether it was due to Pelsaert's description or similar misinterpretations by others, the notion that marsupials were born through the nipple or as a bud from the nipple became commonly accepted and persisted into the 18th and 19th centuries, and was even an article of folk lore into this century. Accompanying this idea was the notion that during reproduction the semen of the male was accepted directly into the pouch of the female, and that the pouch was in fact a second, exterior uterus.

Not all scientists believed in these extreme ideas, and as early as 1698, Tyson, in England, refuted the idea that the pouch was an external uterus. He went on to describe the real uteri and vaginae of the American opossum, and noted the structural differences that distinguish the reproductive tract of marsupials from the other mammals.

With Cook's discovery of the east coast of Australia, it gradually became apparent that there were a great many different species of these strange mammals, and it became increasingly difficult to categorise them as unusual variations of European species. This led eventually to something of a revolution in the classification of mammals. Earlier, mammals had been categorised in terms of general lifestyles and feeding habits. Thus, opossums were put with other sharp-toothed species such as hedgehogs, shrews and pigs, and even late in the 1700s kangaroos were classified as rodents because they had two lower incisors.

It was not until the early 19th century that the marsupials and monotremes were classified separately from the other mammals on the basis of the structure of their reproductive tract and the mode of reproduction and development of the young. This acknowledged that there may be more fundamental distinctions between mammals than their food habits, and it happened largely because a set of species had been discovered in Australia which almost duplicated the well known mammals of Europe except for their novel means of procreation.

The introduction of the Theory of Evolution in the middle of the 19th century was followed by the idea that marsupials represented an intermediate stage along the evolutionary path from egg-laying in reptiles and the egg-laying mammals (monotremes) to giving birth to advanced live young in the placental mammals.

In 1880 the great champion of early evolutionary theory, T.H. Huxley, proposed the terms "Prototheria" (prototype or early mammals) for the monotremes, "Metatheria" (changed or improved mammals) for the marsupials, and "Eutheria" (complete mammals) for the mammals with which the scientific world was most familiar. This terminology was to influence the way scientists interpreted the adaptations of

marsupials for nearly a century.

Prototheria, Metatheria and Eutheria remain in some classifications the names of three sub-classes of mammals, but thankfully their evolutionary overtones are fading.

Scientists now dislike the use of the term "eutherian", but many are equally uncomfortable with the alternative "placental mammal" which was coined at a time when it was wrongly thought that the marsupial foetus was not nourished by a placenta. It has been argued that "placental" has precedence under the rules of scientific nomenclature because it was coined first, and this is the convention adopted here. Therefore we will describe the mammals as monotremes, marsupials and placentals, but it should be emphasised that placental is a term of classification only and not a means of distinguishing between the mammals.

Even into the first half of the present century marsupials were considered to be primitive and inferior anatomically, physiologically and even behaviourally to other mammals; a kind of experiment performed by evolution before everything came together properly in the "true" mammals.

Until recently scientists who studied marsupials did so with the notion that they were looking directly at representatives of the ancestral mammals preserved into the present by isolation from the rest of the world, like the prehistoric monsters of Hollywood movies. Indeed, had a book been published at that time which claimed of any marsupial, as this book will of the

The 19th century naturalist and illustrator John Gould produced many accurate paintings of Australian marsupials, a number of which are now extinct. The broad-faced potoroo depicted here was an inhabitant of southwestern Western Australia, and has not been seen since around 1875.

This delicate marsupial, the desert rat-kangaroo, painted by John Gould in 1843, was last seen in northwestern South Australia in 1931. Some 75 cm from nose to tail, it had an unusual yet very graceful motion, striking the ground obliquely with its left foot some 7 cm in front of the right foot.

koala, that it was highly evolved and efficiently specialised to deal with environmental and nutritional challenges insurmountable to most other mammals, the reaction of scientists would have been scepticism at the very least.

The notion of inferiority or superiority no longer dominates research on marsupials, and has been replaced by a fascination with the remarkably similar evolutionary adaptations of the marsupials and placental mammals. Where there are differences in their physical, physiological and ecological adaptations these are perceived as alternative approaches to meeting similar ecological and environmental challenges.

Since the first marsupials reached Australia a diversification, or radiation, has taken place to produce a wide range of animals suited to many different environments and ecological niches. The similarities between marsupials and placental mammals adapted to similar roles are frequently remarked upon. For example, the teeth and form of the possibly extinct Tasmanian wolf resemble the wolves and other dogs among the placental mammals. The burrowing wombat and the badgers are often compared for their short, stout legs and rotund body. The gliding possums with their membranous "wing" stretched between the fore and hind legs are, externally, almost carbon copies of the flying squirrels of North America, even down to the pattern of stripes on the face and body; and the placental mammal ungulates (such as sheep, cattle, buffalo and antelope) with

their enormous fore-stomach modified to digest plant cell-walls have a parallel in the kangaroos which also have a large and specialised fore-stomach. The list goes on and on, but it is also interesting to consider the types of placental mammals for which there are no marsupial counterparts.

Among the marsupials different species are specialist feeders on small vertebrates, insects, spiders and other small invertebrate animals. There is one specialist anteater, there are several of the possums and gliders which specialise on sap and other exudates from trees, there are species that seek out fungus, and others which specialise in fruit. There are the plant eaters, like koalas, possums and kangaroos, which between them eat a wide range of different types of plants. There are burrowing marsupials, ground-dwellers and tree-dwellers; some are active at dawn and dusk, others intermittently throughout the day and night, and others are only active at night.

But there are no flying marsupials like the bats among the placental mammals; there are no strictly aquatic marsupials like the seals and whales; there are very few semi-aquatic marsupials, or marsupials which specialise in seed-eating, as do many of the placental mammal rodents; only one species lives all of the time underground, and only one marsupial is active only during the day. There are no very large carnivores like the lions and tigers, nor very large plant eaters, like elephants, buffalo and rhinoceros. And there are no species with the extreme development of the

ADAPTING TO A CHANGING WORLD

The fossil record shows that between about 300 and 190 million years ago a line of reptiles split away from the ancestors of the present-day reptiles and the dinosaurs. This evolutionary branch began with the pelycosaurs, which were clearly reptilian, and culminated with a group of mammal-like reptiles, the therapsids. Around 180 million years ago the first mammals had arisen from these reptilian ancestors.

Early in the evolution of the mammals the ancestors of the monotremes split off from the main stream and have evolved separately ever since. It was some time later that the marsupials and the placental mammals began to evolve in different directions. The precise date is a matter for conjecture, but it seems to have been earlier than 80 million years ago.

At one time it was thought that the common features of the monotremes, marsupials and placental mammals were those of the earlier mammals retained by all three groups. But it is now clear that all three groups have changed considerably from their ancestors. They exploited most of the environments present on earth at the time the mammals first appeared, but as the earth and its environments changed, new opportunities arose and the mammals evolved to exploit these.

The fossil record tells us of many changes in the skeleton, especially the skull and teeth, which gave mammals their unique characteristics. It is easy to misinterpret these as advancements or improvements, but it is more correct to regard them as appropriate adaptations for the exploitation of new ecological niches.

It seems that very similar solutions to similar environmental challenges have evolved independently within each group of mammals. This is termed "convergent evolution", and studies of the adaptations of the marsupials and other mammals have given biologists a special insight into this phenomenon.

The reptiles are remarkable animals, but they are only able to exploit a relatively limited range of environments.

The reptilian lower jaw consists of three bones with unspecialised teeth that are sharp pegs. This type of jaw is adapted to exert a limited amount of force and to grasp food and tear it into large pieces. Digestion is therefore slow, and the reptiles are adapted to receive and use food energy slowly, making little use of plants as food.

The reptilian limbs sprawl out to the side and the backbone is capable of little movement other than from side to side, thus restricting their locomotion. They lack an internal mechanism to keep their body temperature constant, and are therefore confined to climates in which they can heat or cool themselves by behavioural means.

The mammalian lower jaw was reduced to a single bone which articulated directly with the upper jaw, and a different muscular configuration gave greater strength and precision of

brain found in the primates. There were large herbivorous and carnivorous marsupials which are now extinct, but other omissions still have to be explained. It has been suggested that marsupials themselves may have been restricted in the scope of their evolutionary adaptations. However, it seems more likely that in some cases suitable niches did not become available during the evolution of the marsupials in Australia, and in other cases many niches were already occupied by other animals (including insects and birds as well as other mammals of the time.)

The Association Between Marsupials and Australia

Today marsupials are found naturally in North and South America, Australia, New Guinea and some islands in the region of New Guinea. By far the largest numbers and variety are found in Australia, the site of their greatest success.

The best explanation for their association with Australia derives from the Theory of Continental Drift, which

movement. The arrangement of teeth became more precise so that upper and lower teeth came into direct contact. A complex pattern of ridges and valleys on the teeth enabled them to grind and cut their food. Front and back teeth differed in shape and size and served different functions. Thus, food could be chopped into smaller pieces, speeding up the digestive process. The precise alignment of teeth was maintained by a system of sequential tooth replacement, but this could only happen a limited number of times rather than indefinitely as in reptiles. The nasal passage also became separated from the mouth cavity so that prolonged chewing was possible.

These changes set the scene for the further specialisation of mammalian teeth into cutters, tearers and large, flat grinders, allowing them to exploit the abundance of flowering plants found in the later evolution of the mammals.

Further adaptations allowed the mammals to become more active. A modified backbone allowed movement in all directions, the legs became closer to the body, permitting greater mobility and manoevrability. New forms of locomotion such as galloping, bounding, hopping and climbing became possible. This, however, reduced their stability, just as the stability of a table is reduced if the legs are brought closer together, so the complexity of the brain, especially those parts concerned with muscular co-ordination increased. Other areas of the brain developed, allowing animals to lead a much more complex lifestyle. For example, the development of the internal regulation of body temperature gave mammals the ability to be active in extremes of temperatures, particularly during the cool of the night.

Nocturnality opened up a whole new world to the mammals, but this would not have been possible without a whole series of adaptations that enabled them to extract and use food energy with greater efficiency than had been seen before.

Associated with these physiological changes were changes in the structure and function of the heart, and an increase in the efficiency of the kidneys and excretory system. External ears developed, increasing the directional discrimination of hearing, and the two bones left over in the transformation from a reptilian to mammalian jaw moved into the middle ear to become two of the three tiny bones that amplify the sound passing through the middle ear, hence contributing to the sensitive hearing of mammals. Specialisations of the skin formed hair, sweat glands, and one of the most significant adaptations of the mammals: the mammary glands.

The female mammal's ability to produce milk to feed her young allows her the freedom to live in a greater range of environments than the reptiles and many other animals which must have access to food suitable for both adults and young. Young mammals have no need for teeth during the rapid growth phase when young reptiles need a succession of unspecialised teeth to keep up with the enlarging jaw. It has been suggested that the freeing of early mammals from this need for teeth early in development made it possible for them to develop their specialised, precisely positioned teeth.

proposes that about 250 million years ago all of the world's land masses were clumped together into a single super-continent known as Pangaea. By the time Pangaea broke up representatives of all of the major groups of animals, including mammals, were present throughout the world. Pangaea split into several large pieces, one of which, Gondwanaland, consisted of the land masses that were to become South America, Africa, India, Australia, New Zealand and Antarctica.

Gondwanaland drifted into the Southern Hemisphere and began to split up. Africa went north and came to rest against southern Europe, India drifted off and collided with Asia forming the Himalayas in the process, and New Zealand floated off into the Pacific by itself. For some time, Austra-lia, South America and Antarctica remained together.

Although we do not know where the marsupials first came into existence, they were once widely distributed in both North and South America and even Europe. They were quite small

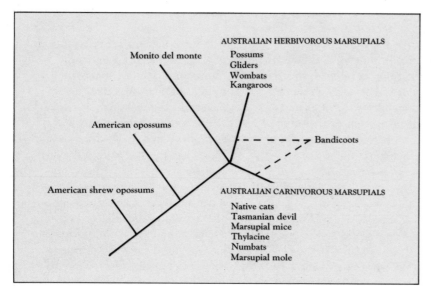

AUSTRALIAN HERBIVOROUS MARSUPIALS
Possums
Gliders
Wombats
Kangaroos

Monito del monte

American opossums

Bandicoots

American shrew opossums

AUSTRALIAN CARNIVOROUS MARSUPIALS
Native cats
Tasmanian devil
Marsupial mice
Thylacine
Numbats
Marsupial mole

This scheme of the evolutionary relationships between the living marsupials shows that the monito del monte (Dromiciops australis), the sole South American member of the otherwise fossil family, the Microbiotheriidae, is the closest relative of the Australian marsupials. Among the Australian marsupials the herbivorous and carnivorous species are seen as quite distinct, but the affinities of the bandicoots remain unresolved.

and, although they outnumbered placental mammals at times and in some places, they were not major animals of the time, for this was the age of the dinosaurs. As the dinosaurs declined so did the marsupials in North America, although it is not known whether the reasons were the same. Marsupials subsequently flourished in South America as small omnivores and carnivores.

It is thought that small, insect-eating representatives of the South American marsupials came to Australia via Antarctica when they were connected by a land bridge and the Antarctic coast was forested, and recent discoveries of marsupial fossils in Antarctica supports this idea. For the following 40 or so million years marsupials had Australia to themselves, and evolved to fill many of the ecological niches occupied by placental mammals elsewhere in the world.

From the small insectivores evolved a range of other meat and insect eating species. With the evolution of the flowering plants came such plant-eaters as the possums and gliders, and the equivalent of the large placental mammal grazers, the kangaroos, some up to three metres tall, and rhinoceros-like four-footed herbivorous marsupials.

The South American marsupials have been greatly reduced in number and diversity largely because they have been displaced over millions of years by other mammals. The Australian marsupials have also faced mass extinctions. Around 20,000 to 30,000 years ago the large herbivores disappeared, due in part to changes in the climate and the coming of mankind. Many small kangaroos and wallabies have become extinct or have been drastically reduced in range and numbers. Others have fallen prey to such efficient hunters as the introduced dogs, cats and foxes.

Types of Marsupials and Classification

The great diversity in the marsupials is now recognised by classifying them as a super order, the Marsupialia, which contains all living members of the mammalian sub-class Metatheria. The Marsupialia is divided into three to five orders depending on which classification you consult. The placental mammals have long been classified into some 18 orders, but it took time for taxonomists to recognise that a lot of fundamental variation also existed among the marsupials.

Below this level the classification of

marsupials is complex and, at present, confused. Everyone can agree on dividing them up into species and even families, but things are a shambles when it comes to higher levels of organisation. This is because so little is known of the ancestors of marsupials, and the fossil record is particularly deficient at critical points. However, there is a measure of agreement about how marsupials fit into *natural* groupings.

The opossums of South America are a group of insectivorous and omnivorous marsupials, most of which live and feed in trees, or at least spend some time in them. These were the first species discovered by Europeans, and they resemble the Australian possums in size and general appearance, but are quite distinct facially.

There are two other groups of South American marsupials. One is represented by a single, mouse-sized, nocturnal, species called *Dromiciops australis*, which lives in the moist forests of Chile and the Argentine Andes. It is an insect-eater, and is of particular interest to Australians because the arrangement of bones in its feet suggest that it may resemble the ancestors of the Australian marsupials. The other group comes under the family name Caenolestidae. They are rat or shrew-like in appearance, and also live in the moist forests of Chile and Argentina. Their food includes small insects, reptiles and birds. They have not been well studied and their relationship to Australian marsupials appears to be relatively distant. The South American marsu-

pials generally either lack a pouch or have a poorly developed one.

The next natural grouping is the carnivorous and insectivorous Australian marsupials. This group includes the various members of the diverse family Dasyuridae and several other families. Among the dasyurids are the spotted quolls (so-called "native cats"), which resemble placental mammal weasels both in diet and general appearance, and have come into competition with introduced cats and foxes as a consequence. There are the marsupial "mice", such as the well-known *Antechinus* species, which resemble mice in appearance, although they are meat and insect eaters. Other members of the Dasyuridae include tiny species like the planigale, which lives in the arid inland and is one of the smallest mammals, weighing only four to five grams, and much larger species like the Tasmanian devil (about 14 kg).

Other carnivorous marsupials are the possibly extinct Tasmanian wolf or thylacine, the marsupial anteater or numbat, currently restricted to small areas of natural habitat in Western Australia, and the blind marsupial mole, found in the sandridge deserts of central Australia, and bearing a remarkable resemblance to the placental mammal mole.

The plant-eating marsupials: the kangaroos, the possums and gliders, the wombat and the koala, are generally considered to be another natural and interrelated group. They probably evolved from an ancestor resembling the small, unspecialised dasyurids.

They range in size from the ten gram honey possum and feather-tail glider to the 70-80 kg male red kangaroo, and they feed on a range of foods from nectar to grass, herbs and tree leaves. The possums and gliders are a diverse group and relationships between species are not clear. The koala is now considered to be more closely related to the wombat than to the possums.

Two families of marsupial have not yet been mentioned. These are the two evolutionary lines of bandicoots, the common bandicoots of the Paramelidae family, and the rabbit-eared bandicoots or bilbies of the Thylacomyidae family. The common bandicoots are like large rats in appearance, and are adapted for digging in forests and grasslands, feeding on mainly small animals dislodged in the process. The bilbies are also diggers, but unlike bandicoots live in burrows and are adapted to life in the arid interior of Australia.

It has long been a problem deciding which group of marsupials the bandicoots belong to. Some systems of classification are on the basis of teeth, in which case the bandicoots are placed with the carnivorous Australian marsupials. Other systems based on foot structure place them with the herbivorous marsupials.

In attempting to work out this and other problems of marsupial classification taxonomists have turned to other features such as blood proteins, chromosomes, and sperm characteristics. At the moment, however, there are more different suggestions than ever about marsupial interrelationships.

What is a Marsupial?

Marsupials get their name from the pouch (marsupium) that covers the nipple area in females, and protects young marsupials during the suckling phase of their development. This characteristic fascinated the first Europeans to discover the marsupials, and it has remained their main distinguishing feature, but it cannot be considered a diagnostic feature because not all marsupials have a pouch. Male marsupials have no pouch (except for one odd South American species in which the scrotum may be drawn into a kind of pouch), females of some species have only a poorly developed pouch, or one which only appears in the breeding season, and some female marsupials have no pouch at all. Although placental mammals do not have pouches, the echidna, a monotreme, does.

So what then distinguishes marsupials from the other mammals? And are marsupials in fact a distinct group? Palaeontologists believe the marsupials have been evolving separately from other mammals for more than 80 million years, so there is little reason to doubt that they are a separate group. In terms of physical characteristics, however, the marsupials are often difficult to distinguish from other mammals because many of their ancestral characteristics have been retained by all mammals, and because the marsupials resemble placental mammals adapted to similar environments.

The way marsupials differ most clearly from all other mammals is in the

MARSUPIALS ARE ALSO PLACENTAL MAMMALS

During pregnancy in humans and many other mammals the membranes of the embryo form an association with the walls of the mother's uterus through which it receives nutrients and oxygen and can dispose of some waste products. This is the placenta, the remains of which constitute the after-birth when babies are born.

In most marsupials only a weak attachment occurs, and the membranes differ in origin from the placenta in other mammals. In a few marsupial species, however, including the koala, the membranes are similar in origin although the attachment is quite weak.

The bandicoots come closest

to the other mammals in this respect. The tissues of the embryo and mother become fused and the barriers between them break down to maximise exchange of nutrients, gases and wastes. In the bandicoots development of the embryo is very rapid and the new-born is more advanced than in other marsupials.

---

structure of their reproductive tract and in the development of their young. In marsupials most of the development of the young from embryo to independence occurs on the mother's teat. Although the monotremes lay eggs they too rely on lactation for most of the development of the young after hatching. In the placental mammals a much larger proportion of development of the young takes place in the uterus, and in some species the offspring is almost independent at birth.

Reproduction in Marsupials

The mammals do not have a monopoly on complex systems of reproduction. In many other animals, including the invertebrates the sexes are also separate, their reproductive systems also include elaborate ducts and accessory organs, and there are equally efficient mechanisms for bringing sperm and ova into contact. In virtually all major groups of animals some species give

birth to living young. The common plan of two testes in males or two ovaries in females linked to ducts which carry the sperm or ova to their respective destinations is repeated over and over again in the animal kingdom. So it is not surprising to find that the basic plan of the reproductive tract in marsupials is the same as that in other mammals.

Marsupial reproduction does, however, differ in two important ways: the young marsupial is born at a very undeveloped stage, roughly equivalent to an early stage placental mammal foetus; and most of its development to independence occurs outside the mother's body.

Female marsupials go through an oestrous cycle similar to the menstrual cycle of humans, lasting between 22 and 60 days, with most being around 25-33 days, coincidently close to the length of the human cycle. During this cycle ova are released from the ovary

THE FEMALE REPRODUCTIVE TRACT IN MARSUPIALS

Many of the organisational differences which were once thought to signify vast differences between marsupials and placental mammals in the process of reproduction itself occurred principally because changes in the "plumbing" system associated with the kidneys were made differently in these two groups. Consequently, there is now much debate about just how significant the differences really are.

To understand how the "plumbing" of the excretory system could influence the organisation of the reproductive tract we must consider briefly the early development of these systems.

In reptiles, birds and mammals the reproductive and urinary tracts develop from three paired sets of tubes, or ducts. These are termed the Mullerian ducts, the Wolffian ducts, and the ureters. In the undeveloped foetus all three sets are present; if it develops into a male the Mullerian ducts degenerate and the Wolffian ducts become the tubes transporting sperm from the testes to the penis (the vasa deferentia); if it develops into a female the Wolffian ducts disappear and the Mullerian ducts become the tubes transporting

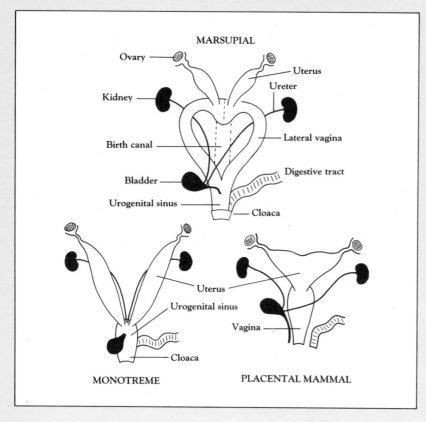

and the reproductive tract goes through the initial stages of preparation for pregnancy. If the ova remain unfertilised the changes are reversed and the cycle starts again.

Only during cycling are females sexually active, and the great majority of marsupial females only cycle during a fixed breeding season. This season is timed so that the offspring leave the mother's pouch when food and climate are favourable. Large species like kangaroos and koalas, in which pouch life occupies the best part of a year, usually breed close to summer so that the young leave the pouch in the following spring. Very small species with a short pouch life often breed in winter. A few marsupials have only one cycle each breeding season, but most have several as long as they do not have a suckling offspring.

A very important characteristic of marsupial reproduction is that pregnancy does not interrupt the oestrous cycle. In most species pregnancy is actually complete before the next cycle is due to begin, and the mother's cycle is only suspended when the new-born marsupial starts suckling.

Marsupial ova are surrounded by a very thin shell which remains in place

ova from the ovaries (the oviducts). Regions of the oviduct also give rise to the uterus (or womb) and the vagina. The ureters are the tubes draining the kidneys and transporting urine to the bladder.

In female reptiles, birds and monotremes the two oviducts remain separate throughout their length, and although there is a region specialised as a uterus (in which the shell of the egg is deposited) there is no vagina. The oviducts and the ureters from the kidney empty into one side of a common tube called a "urogenital sinus", and the entrance to the bladder is on the other side. Thus, the ureters do not empty into the bladder itself, and the urine must pass into the urogenital sinus and then separately into the bladder (not a highly efficient system). The urogenital sinus then meets up with the end of the digestive tract and this common tube, now called a "cloaca" exits to the outside.

Some major changes have occurred in the marsupials and the placental mammals. The oviducts are specialised into not only a uterus but also a vagina, and the ureters empty directly into the bladder. For the ureters to reach the bladder on the other side of the urogenital sinus they move during the development of the young, and they take different paths in marsupials and placental mammals. In placental mammals they move around the oviducts to the outside, while in the marsupials they travel between the oviducts.

In placental mammals there has been in the course of time some fusion of the two oviducts to form one large uterus and a single vagina, the situation with which we are familiar in humans. The marsupials do not have a single uterus, but there is some fusing of the oviducts in the vaginal region. However, because the ureters pass between the oviducts it is not possible for complete fusion of the vaginae to occur. So in marsupials there is a degree of fusion at the top of the vaginal area to form two vaginal culs-de-sac, and these are divided either completely or partially by a membranous wall or septum. From here the vagina divides into two lateral vaginae which then pass into the urogenital sinus.

When female marsupials give birth this difference between the double reproductive tract of the marsupial and the single tract of the placental mammals decreases because a birth canal opens between the culs-de-sac and the urogenital sinus, and birth takes place through this rather than through the lateral vaginae. In the earliest marsupials birth may have occurred via the lateral vaginae, but things have changed so much that in some modern marsupials, notably the kangaroos, the central birth canal remains open and functional throughout life. In others it forms anew at each birth.

The urogenital sinus of marsupials passes into a cloaca which is generally reduced compared with that in monotremes, whereas in the vast majority of placental mammals there is no urogenital sinus and no cloaca, the urinary, reproductive and digestive tracts opening separately to the outside.

for most of the 12-38 days of pregnancy, during which time the development of the embryo is very slow. Towards the end of pregnancy the shell ruptures and the embryo attaches to the wall of the uterus via a type of placenta. Oxygen and nutrients pass from the mother to the foetus, and its growth and development accelerate dramatically. This period of rapid growth lasts only four to ten days, however, and the marsupial is born at a very early stage of development.

To facilitate the journey of the newborn to the pouch kangaroos give birth in a sitting position, while other species lie on their side or face towards the ground with the cloaca raised. The new-born moves unaided from the cloaca to the teat which it will not release at all for some weeks. Although the new-born of all marsupials weigh less than one gram, the forelimbs are well developed with claws to assist in the journey to the pouch. The digestive tract and lungs are functional, and the nostrils have smell receptors. There is a primitive kidney and urinary system necessary for excretion of waste. However, in most other respects development has only just begun. The hindlimbs are little more than buds, the eyes are not open or functional, the ears are barely visible, the brain is undeveloped except for areas concerned with smell and the coordination of the forelimbs, and the new-born is not capable of regulating its own body temperature.

The number of young varies from one in the larger species to as many as 18 in small species. Where many young are born there are often not enough teats for them all and the surplus die, or there may be losses of young from the teats. The number of teats differs greatly between species. In the small carnivorous marsupials there are usually between six and ten, and the number may vary within a species. The larger species have up to four teats.

The pouch itself is even more variable. In some species it is non-existent, in others a fold of skin develops on either side of the teat area during breeding to protect the attached young. Some pouches have a very wide opening and don't completely enclose the teat area; and then there are the deep, complete pouches typical of the hoppers, the diggers and the active climbers. In these species the pouch may open forward, as in the kangaroos and possums, or backward, as in wombats and koalas.

Males of some species also go through seasonal cycles of sperm production and testis enlargement, whilst others produce sperm continuously. In both types the prostate becomes up to 100 times larger during the breeding season, permitting the production of relatively vast quantities of semen at these times. Male marsupials are usually promiscuous, and the capacity to produce a lot of semen allows them to mate with several females and hence maximise their chances of having offspring. In copulation with the female the sperm move rapidly up the lateral vaginae to the uterus where fertilisation takes place before the shell membrane has been deposited on the egg. Although many sperm lodge in the outer layer of the egg, only one penetrates to the centre.

A particularly interesting variation on the marsupial pattern of reproduction is seen in the kangaroos. This is known as "embryonic diapause", and as the name suggests, it involves a pause in the development of the embryo. In kangaroos the mother mates immediately after giving birth. But as long as she has a pouch young which is suckling vigorously, the new embryo only develops to a very early stage, and it remains suspended in this state until either the mother loses her pouch young through misadventure or its rate of suckling declines as it is weaned. At this point the delayed embryo resumes development and the mother will give birth after the usual period of gestation. This adaptation allows rapid replacement of lost young and reduces the time between weaning of one young and the birth of another. However, it is important to note that this is not a feature of all marsupials as is often thought, and that it also occurs in some placental mammals.

Lactation in marsupials is quite complex because the nutritional require-

THE MALE REPRODUCTIVE TRACT IN MARSUPIALS

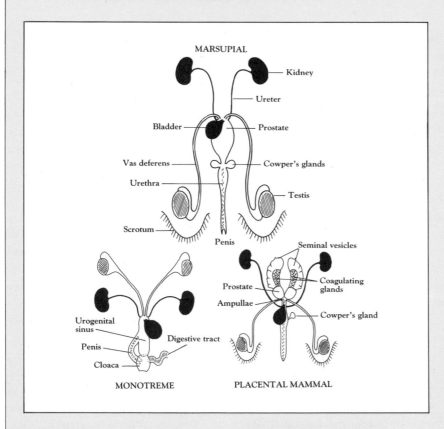

The reproductive tract of male marsupials is very similar to that of placental mammals but is less similar to that in monotremes. As in most placental mammals, the testes of most marsupials are housed in a scrotum, or sack, which hangs outside the body where it can be cooled in the wind. This is important because even the temperature that the centre of the body rises to in exercise can inhibit the production and maturation of sperm.

In both marsupials and placental mammals there is a similar network of blood vessels (the rete mirabile) in the neck of the scrotum which regulates the temperature of blood passing to the testes. Interestingly, it is suggested that both the scrotum and the rete mirabile evolved separately in the two groups.

From the testes run the vasa deferentia, the tubes formed from the Wolffian ducts, which carry sperm from the testes to the penis. These run into the urethra, the common tube leading to the outside via the penis, which carries both semen during copulation and urine during urination.

But there are some differences between male marsupials and male placental mammals. The arrangement of urinary and reproductive ducts differs just as it does in females. Thus the ureters pass between the vasa deferentia in male marsupials whilst they pass around in male placental mammals. In placental mammals there are several groups of accessory organs which provide most of the constituents of semen other than sperm. These are the prostate, the Cowper's glands, the seminal vesicles, ampullae and coagulating glands.

In marsupials only the prostate, which is enlarged to take on a greater responsibility, and Cowper's glands are present. The penis of most marsupials is bifurcate (forked), but in kangaroos it tapers to a single point. It is not forked in placental mammals. It was once thought that the two heads of the marsupial penis were to direct semen into the two lateral vaginae of the female. However, kangaroos do not have a forked penis yet semen gets into and through both vaginae, so in this case two heads are not necessarily better than one!

Because marsupials have a cloaca, the penis is withdrawn into the common opening when not erect, and the penis is thus situated behind the scrotum rather than in front of it as in most placental mammals.

A further interesting distinction is that male marsupials never have nipples or mammary glands, whereas nipples and the associated gland are common in male placental mammals although seldom functional. It has been suggested that retention of the mammary gland in male placental mammals may be of benefit because it is conceivable that the male could help in feeding the young if the female is unable to (a few such cases have been recorded). In marsupials this is impossible because the baby is firmly fixed to the nipple during the early part of lactation, and in fact male marsupials play a very minor role in general in raising their offspring.

ments of the young change dramatically during their development. To cope with this the mother produces essentially two types of milk, and in marsupials like the kangaroos, where two young of different ages may be suckling at the same time, the mother produces two types of milk simultaneously.

During early lactation the suckling young undergo changes similar to those of a placental mammal embryo in the uterus, developing organs, skeletal features, and physiological capabilities that will characterise them as adults. The milk at this stage is high in carbohydrate to supply energy, low in fat which cannot be handled by the underdeveloped digestive system, and contains all the necessary proteins.

In late lactation the young marsupial grows rapidly and thus needs more energy at a faster rate. This is the time when kangaroo joeys are seen hanging out of the mother's pouch and making brief excursions on their own. The milk at this stage is high in fat because this supplies more energy for the same weight or volume of milk, and low in carbohydrate. The concentration of protein in the milk rises slightly and the constituent parts of the protein, the amino acids, change in accordance with the requirements of the young. The thickness of the milk also increases from about the consistency of cow or human milk in early lactation to up to four times as thick in late lactation.

It is becoming clear that the hormonal control of the switch from one type of milk to the other in marsupials is quite complex and that no comparable situation arises in placental mammals. The ability of some kangaroos to make both milks in different mammary glands at the same time amazes and baffles those who study lactation in placental mammals.

As the young marsupial grows, so its need for milk increases. To meet these requirements the mammary gland grows in size throughout lactation, reaching a peak shortly after the change in milk composition has been completed. As the young is weaned the gland gradually regresses to be ready to start a new cycle.

The larger marsupials like koalas, some possums, wombats and kangaroos, with a complete pouch, carry their single young around in the pouch until it is able to keep up with the mother or ride on her back. The smaller species with a poorly developed pouch and large litters carry the young around until they get too heavy and are no longer permanently attached to the teat. They are left in a nest while the mother goes away to feed. Those species with a complete pouch but with two to four young also leave them in a nest in the later stages of lactation.

The young of small marsupials are weaned quickly after two to four months of lactation. In large species the young may leave the pouch permanently after seven to ten months although they still take milk intermittently until they are 12-18 months old. Some of the most remarkable lactators among marsupials are the bandicoots which develop faster than other species during pregnancy. For them lactation lasts about two months during which time an average of three young weighing between them half the mother's own weight are raised and weaned.

Marsupial Body Plan

Marsupial skulls generally have a small brain case and a larger face area than the placental mammals. The palate usually has holes in it between the upper molars, and the rear part of the lower jaw is usually turned inwards, unlike the lower jaw of placental mammals. The eye socket merges with the opening through which the muscles attaching to the lower jaw pass, and other small differences include the site at which veins or nerves pass through the skull and the position of ridges and thickenings for muscle attachment.

Many marsupials have more teeth than the placental mammals. The small to moderately large insect and meat-eating marsupials of South America and Australia have relatively simple, sharply pointed teeth used for gripping, tearing and macerating. This is considered to resemble the state of dentition of the earliest mammals. Starting from the front of the jaw, there are usually four or five pairs of narrow, pointed upper incisors and three lower pairs; the canines are well developed in both upper and lower jaws; there are two or three pairs of blade-like premolars in each jaw and four pairs of molars bearing sharp, shearing cusps. This adds up to 48 teeth, although American opossums have 50. Comparable placental mammals have fewer upper incisors, one more premolar but one less molar,

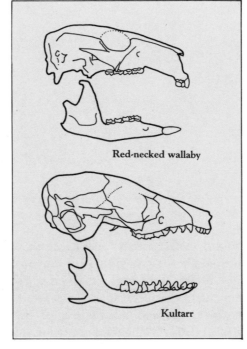

Red-necked wallaby

Kultarr

a total of 44 teeth.

Evolution of specialised teeth for eating plants has meant changes to and the loss of some teeth. The herbivorous kangaroos, possums, wombats and koalas have only one pair of lower incisors and up to three pairs of upper incisors. The lower incisors have developed as large, forward-facing teeth useful in precisely choosing an individual leaf or blade of grass, and the neatness with which these animals nip off their chosen item of food is quite remarkable. The canines are reduced and sometimes absent, and the premolars are often reduced to only one pair. This leaves a space between the front and back teeth called a "diastema" which allows herbivores to move the food

Marsupial skulls generally have a small brain case and large face area, with holes in the palate between the upper molars, ridges and thickenings for muscle attachment, and the eye socket merged with the opening through which the temporal muscles attached to the lower jaw pass. The red-necked wallaby is a herbivorous diprotodontid marsupial, having only two large and forward-pointing lower incisors. The large gap between front and rear teeth facilitates the movement of grass and leaves around the mouth; the molars are flattened for grinding. The mouse-sized kultarr is a carnivorous marsupial, feeding on insects and small vertebrates. It has six lower incisors, and is classified as a polyprotodont (having four or more lower incisors), the teeth are sharp and adapted for cutting, tearing and macerating.

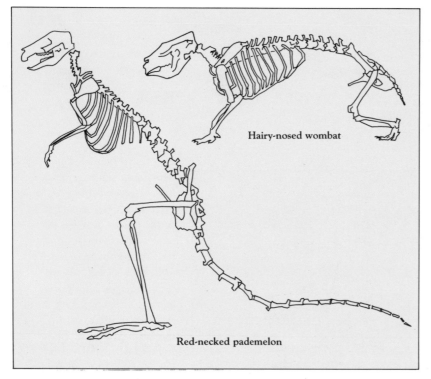

Hairy-nosed wombat

Red-necked pademelon

Skeletons of the red-necked pademelon and hairy-nosed wombat clearly indicate modifications to the basic mammalian form that has enabled these two marsupial species to occupy quite different ecological niches. The red-necked pademelon has small forelimbs and large hindlimbs for leaping through the undergrowth. The hindfoot is narrowed and lengthened for thrust and balance, the tail is extremely long and is used as a third prop, facilitating their upright posture. The hairy-nosed wombat is a burrowing herbivore, and is perfectly adapted for digging large, complex burrows, with short legs, powerful shoulders, short tail and long flat claws. Like other marsupials both possess long epipubic bones that project forward from the pelvis.

around in the mouth and to chew it thoroughly without biting the tongue. The molars are large, flat, grinding implements with considerable variation in the shape and positioning of the cusps, relating to the type of food eaten and its mechanical and abrasive properties in particular.

Unlike most placental mammals which have a complete second set of teeth, marsupials only replace the first premolar.

Marsupials with four or more lower incisors are known as "polyprotodonts", and those with two lower incisors are termed "diprotodonts".

The skeleton of marsupials is essentially mammalian and few unique features are present in all species. There are the usual array of modifications to bones to make them more efficient for digging, climbing, running, jumping and hopping. The most distinctive feature of the skeleton of marsupials is the presence of two small bones projecting from the pelvis forward into the wall of the belly. These are known as

the "epipubic" bones, and it was once thought that they support the pouch. However, they are present in both sexes and in monotremes and reptiles, although not in placental mammals. Their precise function is not clear.

The hands of marsupials do not vary dramatically between species. The five digits are all recognisable, although the first and/or fifth are reduced in size in some species and do not always bear claws, but the feet show dramatic variation, and like the variation in dentition has been used to classify them. The feet and hands of all mammals are based on a common plan of five digits, although many have either lost digits or have two or more fused.

In all the herbivorous marsupials the first digit of the foot is either reduced to a clawless thumb which is opposable to the other digits (as in possums and the koala) or is absent altogether (as in the kangaroos). In the kangaroos the fourth digit is elongated for hopping, and the foot in general is long and thin. The second and third digits of all herbivorous marsupials are reduced in size and connected by a web of skin to form, essentially, a single digit with two claws. This condition is known as "syndactyly". The combined digit is used for grooming, but whether it evolved for this purpose or as a stage towards reduction in the number of toes to provide a more rigid foot for hopping is still debated. The alternative to syndactyly is "didactyly" in which all the digits are free, as in the carnivorous marsupials.

Responding to the Outside World

The brain of marsupials is said to be smaller in relation to body size than in placental mammals. However, this is not to say that all marsupials have a smaller brain than all placental mammals: in both groups there is considerable variation. The neocortex, that region of the brain thought to be

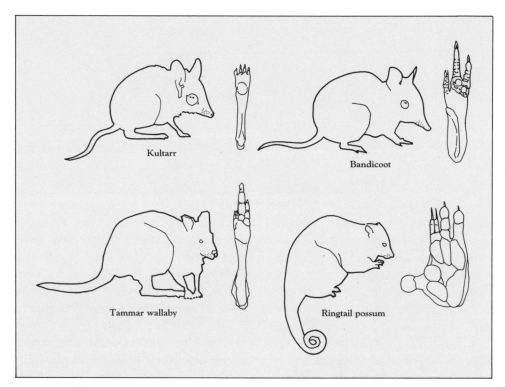

Kultarr

Bandicoot

Tammar wallaby

Ringtail possum

Marsupial feet show a variety of forms. Carnivorous species such as the kultarr have separate toes (didactylous), whereas the herbivores, such as the tammar wallaby and ringtail possum, have the second and third toes enclosed in a web of skin (syndactylous), forming a grooming "comb". Bandicoots have a mixed diet of insects, small vertebrates and plants. Their feet are syndactylous with a much reduced first digit. Ringtail possums, on the other hand, have an opposable first digit and sharp claws for gripping tree branches.

responsible for intelligence and advanced social behaviour, is well developed, particularly in the larger species such as the wombats and kangaroos, which are on a par with placental mammal cats. There are, however, no marsupial counterparts of the placental mammal primates, whales and dolphins, which have a very highly developed neocortex.

The question of the relative intelligence of marsupials has come up time and time again, and has yet to be resolved. In some tests marsupials do better than their placental mammal counterparts, while in others they do worse.

One major structure absent from the brain of marsupials is the corpus callosum, which transfers information between the two halves of the brain in placental mammals. Initial thought that this might impair information transfer in the marsupial brain now seems to be unfounded, and other structures appear to take over the role of the corpus callosum in marsupials.

The majority of marsupials are nocturnal, and their principal modes of communication are sound and scent. Marsupials are not obviously vocal like lions, wolves, elephants, sheep and cattle. Nevertheless, they do have an extensive repertoire of low intensity hisses, grunts, growls, churrs and screeches for antagonistic situations; clicks and clucks for use during sexual arousal; and coughs, squeaks, squeals and zook-zooks for use by distressed youngsters. The possums and gliders, have the greatest vocal repertoires.

Scents are produced not only in urine, faeces, saliva and sweat glands, but also in specialised scent glands in the mouth, pouch, around the opening of the cloaca, and on the chest in some species. Scent marking conveys such information as ownership, dominance, mood and sexual state. The part of the brain given over to the processing of information coming from the nose (the olfactory bulb) is very prominent in marsupials.

The majority of marsupials are either

solitary or live in loosely defined social groups, rarely of fixed composition. Most males are promiscuous, only a few species have a monogamous mating system, and long term bonding between males and females is rare.

Marsupials have little need for complex social systems and behaviour patterns, for, unlike the other mammals they do not hunt cooperatively, like the lions and wolves, and there are no large herbivores active during the day which rely on their social system for protection against predators. Males have little to do with raising their young because of the highly developed relationship between the young and the mother, and it has been argued that a male stands a better chance of passing on his genes to future generations by mating with many females.

Digestive Systems

Just as they have evolved a range of adaptations of the teeth to deal with different diets, so marsupials have developed appropriate digestive tracts, ranging from the relatively simple type found in the carnivores with a small stomach and intestines without elaborate specialisations, through slightly more complex tracts of the omnivores, to the highly specialised digestive systems found in the herbivores.

There are basically two types of digestive tract among herbivorous marsupials with a great amount of variation within each type. In wombats, possums and gliders, and the koala, the region of the gut immediately before the rectum is elaborately expanded in various ways to house micro-organisms which help in the digestion of plant material. This is usually in the form of a large outgrowth from the main large intestine called the "caecum", which corresponds to the region of the human appendix. In the kangaroos a part of the stomach is also expanded to house micro-organisms to help in the digestion of plants, and is the type of adaptation seen in sheep and cattle.

So it is that marsupials have many characteristics in common with other mammals, but they also have their own unique ways of dealing with the Australian environment. They are not the primitive, inefficient, evolutionary throwbacks they were once thought to be; and one particular marsupial has reached heights of adaptation matched by few other mammals in the world; that animal is the koala, to which this book is dedicated.

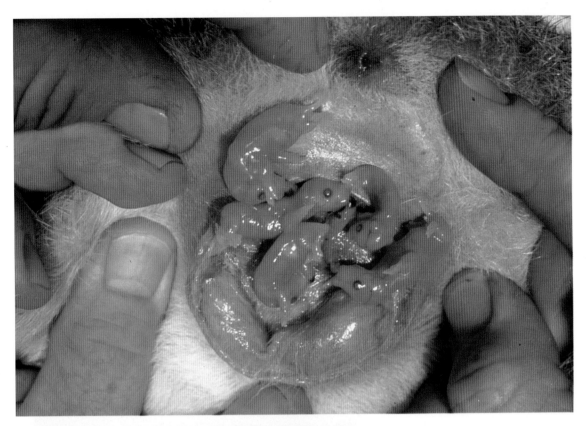

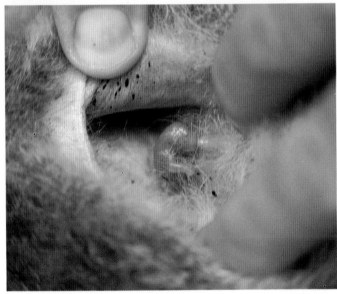

Marsupials are born at a very
early stage of development with only
the forelimbs, digestive and
excretory systems and the smell
receptors functional. The
photograph above shows newborn
bandicoots attached to the teats
inside the mother's pouch, left shows
a newborn koala.

Koalas are one of the largest arboreal animals, and have struck a delicate balance between the minimum size needed to exist on a nutritionally poor diet of eucalypt leaves, and the maximum size it can reach without crashing to the ground in its search for food.

FORM AND FUNCTION IN THE KOALA

"Whether we consider the uncouth and remarkable form of its body, which is particularly awkward and unwieldy, or its strange physiognomy and manner of living, we are at a loss to imagine for what particular scale of usefulness or happiness such an animal could by the great Author of Nature possibly be destined."

The above comments by George Perry appeared in a British natural history periodical of 1810. More recent authors have commented on Perry's tendency towards fancy, but there is no denying that many of the koala's physical characteristics appear at first to be anomalous and mismatched to its arboreal habitat. One has only to observe a koala in captivity for a few minutes (assuming that it is awake and active) to see that this is no aerial gymnast. Its movements are usually tentative and cautious, and particularly on awakening, are punctuated with slips and wobbles reminiscent of a drunken man walking a tightrope. Autopsies of wild koalas, in fact, frequently show evidence of old bone fractures which could only have been sustained in a fall from a tree!

Several of the koala's physical characteristics do, in fact, suggest that it is a ground-dwelling species, possibly related to the wombats, which has adapted to tree life, rather than evolving primarily as a tree-dweller.

But it is unfair to present a picture of an animal which is not at all adapted to its environment. The koala comes into its own when its physiological adaptations are considered. This is a species that has overcome constraints imposed by a harsh environment and an even harsher food source, to achieve a degree of utilisation of tree foliage as food that few species of mammals anywhere in the world even approach. Ironically, perhaps, those very characteristics which are not ideal for locomotion in trees, its large size and sleepy disposition, facilitate the koala's ability to exploit this food source.

The koala is delicately balanced between, on the one hand, the minimum size enabling a mammal to live on nutritionally poor tree leaves, and on the other, the maximum size it can attain and still have enough mobility in the trees to actually gather the leaves. Under these circumstances its slow, energy-conserving movements, may be seen as far more than an appealing attribute that amuses tourists.

Size and Shape

The body proportions of the koala, like so many of its other characteristics, set it apart from other tree-dwelling animals. The koala is short and stocky with remarkably long arms and legs. Many other arboreal mammals, such as monkeys and apes, have both long limbs and body, giving them agility and manoeuvrability when scrambling and swinging through the trees.

The koala, however, is about as large as an arboreal animal can be without falling from the tree, and it needs those long limbs to enable it to leap in an upright position and grip a thick branch or trunk with all four limbs at the same time. Compared to other arboreal marsupials such as the possums, koalas are slow, laborious climbers. Possums are light enough to simply scamper up the trunk almost as if running on flat ground, whereas the koala levers its greater weight up the trunk by first gripping firmly with both front limbs and then moving the hind limbs up to meet them in a bounding motion, similar to the way humans scale coconut palms.

Arboreal mammals range in weight from about 1-15 kg, although few are heavier than 9 kg. This relatively narrow range of body weights in the tree-dwelling mammals reflects the weight that tree branches can support. Heavier animals do not have the necessary mobility to reach fruits, flowers or the most succulent young leaves growing on the tips of branches. Mature koalas weigh from about 4-14 kg, making them larger than most other arboreal mammals. Male koalas are up to 50 percent heavier than females, and both sexes in the southern part of the species' range (Victoria, South Australia) are considerably larger than those in the northern part (Queensland).

Such latitudinal variations are common among species of mammal around the world, those living in warm environments are generally smaller than other members of the same species in

cooler climates. Small animals have a larger surface area in relation to their body size, allowing them to lose more heat through the skin. This gives them an advantage in warm climates, and the opposite is true of larger animals in cool climates.

The koala's feet and hands are endowed with long, pointed claws, of great benefit in the gripping of branches and tree trunks. Anyone who has ever handled koalas will know that their first instinct when removed from the security of being wedged into the fork of two branches is to dig their claws into anything to get a grip, and hence, presumably, to stop themselves from falling.

The hand is well developed as a grasping device, with five similar-sized digits. The first and second are opposable to the other three, providing a powerful pincer grip. This type of hand is termed "forcipate", and is unusual among mammals, although it is found in several of the other arboreal marsupials. It does not give the same degree of dexterity enjoyed by the primates which are able to oppose the thumb individually to each of the other four digits, but it nevertheless enables the koala to grip quite small branches, manoeuvre them to its mouth and choose particular leaves from those available.

The palms of the hands and feet of many mammals possess tiny ridges of skin which give extra grip. Our fingerprints are one example of these "friction ridges", and they are the culmination of a trend towards increasing complexity in the pattern of palm ridges from the primitive primates, through the monkeys, to humans. Such trends are evident in other groups of mammals and in these too are associated with increasing complexity of use of the hands and feet. Possums and gliders show the same evolutionary trend, with the basic pattern of pads

The hand and foot of the koala show a relatively simple pattern of pads and ridges, similar to those of ground-dwelling marsupials such as wombats. The hand, however, is a powerful grasping device, with the first two digits opposing the other three. The first digit of the foot has no claw, and is also able to oppose the others, while the second and third digits are partially fused and used for grooming. Note the long pointed claws, used for extra grip when climbing.

a   b

being elaborated upon and overlaid by variously complicated patterns of ridges. Like many of the primates, these marsupials have evolved to move around in trees where good grip is vital.

The palm of the koala, however, has a very simple pattern of pads bearing no ridges. This is more the pattern expected in a ground-dweller, and is very similar to that seen in other terrestrial marsupials including wombats. It is just one of several characteristics which suggest that koalas did not evolve primarily for life in the trees.

The feet have the same arrangement of pads without ridges, but the digits are different and specialised. The first digit has no claw and is separated from and opposable to the others. These are shorter than the digits of the hand, and two of them are partially fused, separating only at the tip. These "syndactylus" toes are important in grooming, being used like a comb.

The koala's tail is reduced to a stump, and is quite unusual for a tree-dwelling mammal. Arboreal mammals generally use the tail as a fifth limb, wrapping it around branches to give then extra mobility and manoeuvrability, and in some cases using it to carry food and other materials. The koala's tail stump consists of 6-7 caudal vertebrae, compared with 12-13 in the ground-dwelling common wombat, and 20-30 in possums and gliders.

Female koalas have a marsupium, or pouch, with two teats. The pouch opens towards the rear rather than towards the head as in kangaroos and possums, and in this respect the koala

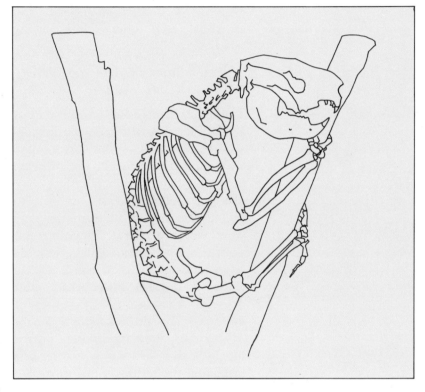

again resembles the wombat. One can envisage many advantages of a backward-opening pouch for a burrowing animal like the wombat, which moves forward with its belly sometimes scraping the ground, but few for an animal that sits upright in a tree. The distance over which the newly born koala has to travel to reach the pouch is reduced, admittedly, but it then faces the possibility of falling out again in later life. The problem is not really that great in early pouch life because the baby is firmly attached to the teat, but as the baby grows the danger of falling from the pouch increases, and one koala at Taronga Zoo in Sydney had to be

The skeleton of the koala resembles that of the ground-dwelling marsupials. The tail is reduced to a stump consisting of only 6-7 caudal vertebrae, and the body is short and stocky. Yet the koala has long limbs and grasping hands, characteristics typical of arboreal mammals, enabling it to move easily between branches.

clipped into its mother's pouch because her pouch muscles periodically relaxed and the baby became airborne!

The similarities between koalas and wombats in the hands and feet, the lack of a tail, the backward-opening pouch in females, and in males, the structure of the sperm, have prompted the suggestion that these two species evolved from a common ancestor, and are today each other's closest living relative.

Temperature Control and Water Conservation

The ability to maintain the body core at a relatively constant temperature, despite fluctuations in the surrounding environment, is a fundamental feature of the biology of mammals. The complex metabolic processes essential to the functioning of all mammals only work within a narrow range of temperatures. The koala maintains its body temperature at about 36 degrees Centigrade, 2 degrees lower than most placental mammals, reflecting a lower rate of metabolism in koalas.

The short, fine and densely matted fur of koalas has some of the best insulating properties found in marsupials, verging on that of some arctic mammals. It differs in colour between animals from different locations, being usually light to dark grey on the back, although sometimes showing touches of brown, with white or yellowish fur on the underbelly. Patches of white are common on the rump.

The fur is thickest on the back and densely matted, it is disturbed little by wind, and so keeps its insulating qualities. Thus on cold days koalas sit huddled in a ball, reducing the surface area through which they lose heat, and exposing the darker fur on the back to absorb as much heat as possible from the sun. On a hot day koalas expose more of the white belly fur which reflects heat and is easily ruffled by the slightest breeze. The pose adopted on hot days resembles that of a drunk, with arms and legs dangled loosely around whatever branches are convenient, and often leaning back to maximise the surface area through which they lose heat. It is this characteristic pose which contributed to the myth that they become intoxicated on gum leaves!

In summer koalas can tolerate temperatures down to 15 degrees Centigrade without any effort, and in winter this lower critical temperature is about ten degrees. Unlike some mammals like seals and bears, koalas do not rely on a layer of fat beneath the surface of the skin as a major means of insulation, and special measures must therefore be taken to maintain the body temperature in cold conditions. Blood flow to the extremities is initially reduced to prevent heat loss through the skin. This technique is also used in humans, giving rise to cold feet and hands in order to maintain the vital organs at the correct temperature, and in extreme cases may lead to the loss of fingers and toes due to frostbite.

If redirection of blood flow is not sufficient then extra heat must be produced by the burning of fuel by rapid contraction of certain muscles, and this is the function of shivering. The extent to which koalas shiver in the wild is not known, but as they live in areas where winter temperatures fall well below ten degrees Centigrade, shivering probably becomes an important temperature control factor at times.

When the air temperature rises above 25-30 degrees Centigrade koalas use their lungs and breathing passages like an air-conditioner to keep cool. The surfaces of these passages are constantly kept moist, and rapid breathing increases the rate of evaporation of this moisture in the exhaled air, drawing heat from the body in the process. Koalas only have sweat glands in the palms of the feet and hands, and sweating in these areas may be more to give grip than to help dump excess heat.

The technique of evaporative cooling is taken further by some animals, notably kangaroos, which spread saliva on bare parts of the skin, especially the forearms, where there are highly branched networks of blood vessels. The saliva evaporates off this surface, taking heat from the blood with it. The extent to which koalas do this is quite small. At very high temperatures and humidity koalas occasionally lick their paws and spread saliva on the face, but this probably has only a minor influence on body temperature.

To overcome the extra water loss through evaporation on hot days, koalas remove as much water as possible from the faeces and urine before excretion, and except perhaps on the hottest days they are still able to obtain all their water requirements from their diet of eucalypt leaves.

The Senses

In the head of the koala are found some of its most distinctive features. The skull itself is large in comparison with the rest of its body, and together with its broad, flat appearance, gives the animal much of its character and has led to the comparison with bears. Surprisingly, in this large skull is housed one of the smallest brains relative to the size of the body of any mammal, but what this means about the relative intelligence of the creature is a matter not yet addressed by scientists.

The vibrissae (sensory hairs or whiskers) on the face are much less numerous, cover fewer areas and are shorter than those on the faces of the other arboreal marsupials sharing the forests with the koala, the possums and gliders. Unlike these forest dwellers, there are no signs of sensory hairs on the palms of the feet and hands of koalas, and none on the limbs generally. In this way the koala resembles the tree kangaroo, *Dendrolagus* among the marsupials, and other slow movers among the placental mammals such as the sloth *Bradypus* and the potto *Perodictus* (a relative of the monkeys) which do not require the same degree of sensory information as fast moving, highly manouevrable species.

Hearing is a particularly important sense to nocturnal mammals and those living in forests where it is difficult to see long distances, and sound is a major means of communication in such a habitat. It is not surprising to find that the possums and gliders rely heavily upon vocal communication and have some of the richest vocal repertoires of all marsupials. These animals are very sensitive to sound and react quickly to strange noises, and their hearing is an important means of defence against surprise attack by predators. Consequently the ears of most of the possums and gliders are large, mobile and free of hair around the entrance and the inside.

Although the hearing ability of koalas has not been specifically studied, it is probably less acute than that of the possums and gliders. The ears of koalas are relatively small compared with the size of the head, and the entrance is often covered with thick fur. Their vocal repertoire is relatively restricted; they are less gregarious than many of the possums and gliders, and so probably have less need for the subtleties of communication employed by these other species. Koalas seem to react less to strange noises, probably relating to the fact that they have few natural predators in the trees, although on the ground they are much more wary, being prey to dingoes and other wild dogs.

Lieutenant-Governor Colonel William Paterson, in his writings of the early nineteenth century described the

EUCALYPT LEAVES: A VERY UNPLEASANT DIET

The chemicals contained in *Eucalyptus* leaves were a topic of great interest even in the earliest days of settlement in Australia. The volatile or essential oils attracted particular attention. The term "essential" refers to the contribution of certain oils (mainly terpenoids) to the smell or "essence" of plants, and the essential oils of eucalypts give Australian forests their distinctive aroma. In fact, one of the first exports from the colony of New South Wales in 1788 was a sample of *Eucalyptus* oil, and in 1852 the essential oil industry of Australia began. At first the oils were used in various medicines for the treatment of conditions like cholic, but more recently they have become important ingredients of industrial disinfectants, deodorants and perfumes.

It was soon realised that essential oils are toxic to humans and other animals when taken in any but very small doses. Koalas were known to eat little more than eucalypt foliage, and biologists began to question not just the toxic effects of a diet of eucalypt leaves, but the overall nutritional value of such a fibrous and unappetising leaf.

On the a positive side, however, eucalypt foliage does have a relatively high water content. The mature leaves contain 40-50 percent water, and the young leaves 50-60 percent. Koalas obtain their total water requirements from the leaves themselves and from rain and morning dew settling on the leaves. Except in severe droughts they therefore have no need to leave their leafy perch to drink.

The dry matter of the eucalypt leaves, however, contains an array of unappealing and strange compounds sufficient to deter most other herbivores.

Carbohydrates, proteins and fats are an essential part of the diet of all animals. They are the fuel supply for the metabolic processes and the building blocks of living matter.

Human diets are traditionally very high in carbohydrate content, particularly sugars and starch, which are easily digested and provide a fast, readily available supply of energy. Yet in eucalypt leaves sugar and starch comprise less than 10 percent of the dry matter, and this low supply of energy-rich food has been thought to force the koala to lead a very quiet life.

Plant tissues contain other carbohydrates known collectively as "fibre". The highest fibre diets eaten by humans contain no more than ten percent fibre, but the dry matter of *Eucalyptus* leaves contains 30-50 percent fibre. Unlike sugar and starch, fibre is difficult to digest and provides little energy, but it has important physical and chemical effects on the digestive process.

One of the benefits of fibre to humans is that it absorbs water in the digestive tract and swells, thereby occupying space and limiting the amount of extra food that can be eaten. Thus fibre is an aid in weight control. But this can go too far, and the fibre content of *Eucalyptus* foliage is so high that the koala could have trouble eating enough to meet its energy needs.

Fats account for about 17 percent of the dry matter of the

koala's eyes as "generally ruminating, sometimes fiery and menacing". Few who have locked gazes with a koala in an animal park could imagine what he meant by "fiery and menacing". However, it is much less of a mystery to those who have witnessed a dominant male "explaining" to a subordinate who ventures into his tree that really he ought not to be there; and would be an appropriate description by those who have attempted to handle a wild koala in a way that gives it anything near an even chance of getting free. But for most of the time the description "ruminating" is quite apt.

Studies of the visual abilities of koalas and the structure and function of their eyes have yet to be undertaken. Externally their eyes resemble those of a cat, with a vertically slitted iris able to rapidly accommodate to low light intensities, expanding the size of the pupil to allow the maximum amount of available light to enter the eye, giving good night vision. Since the koala is

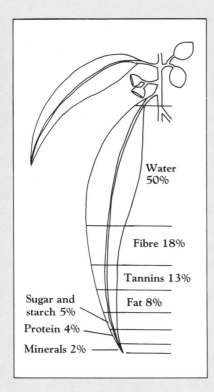

Water 50%

Fibre 18%

Tannins 13%

Fat 8%

Sugar and starch 5%

Protein 4%

Minerals 2%

The importance of protein in nutrition is well known. It is a vital structural component of all animal tissues, being continually broken down and resynthesised by the body. It may come as no surprise to learn that *Eucalyptus* leaves contain relatively little protein, much less than would be needed by most animals to survive, let alone reproduce. The actual amount varies between 5 and 15 percent of the leaf dry matter.

But the koala's problems do not stop here. Another product for which eucalypts are exploited commercially is tannin. Tannins are a group of compounds containing phenol, and are used to toughen and protect leather. They do this by binding to the proteins of the leather and changing their properties. The level of tannins in eucalypt leaves may reach 25-30 percent of the dry matter, a phenomenally high concentration. They have the effect of binding to proteins in the diet, to enzymes in the digestive tract and even to some carbohydrates. Hence tannins have the potential not only to reduce the availability of the already small amount of

protein in the eucalypt leaf, but also to interfere with the digestive process itself.

There is a further problem associated with tannins. Some are broken down in the digestive tract, and this releases their major constituent, phenol. Phenol is found in many disinfectants, and is one reason for their poison warnings. Like the essential oils this phenol is absorbed by the koala and must be dealt with as a toxic substance.

Little is known about the vitamin content of eucalypt foliage or the koala's requirements for vitamins. It is known, however, that eucalypts generally contain high concentrations of ascorbic acid (vitamin C), but the significance of this to koalas is not known.

Similarly, although we know that the concentrations of many of the minerals needed by humans are low in eucalypt foliage, we know little of the koala's need for them.

Faced with such an unappealing, toxic and nutritionally poor diet, it is hardly surprising to find that the koala has the eucalypt foliage largely to itself.

eucalypt leaf, and although this is not high in comparison with a potato chip, it is remarkably high for a leaf. Some of this fat is the toxic essential oil and another large portion comprises waxy compounds. Apart from tasting pretty awful, waxes are not digested to any great extent by animals, and hence give little or no energy.

active at night, their eyes are probably similar in structure to those of other nocturnal marsupials, being specially developed to detect low levels of light. This is achieved by a reflective layer at the back of the eye called the "tapetum lucidum". Light entering the eye passes twice through the light-detecting cells situated in front of this reflecting layer, once on its first passage through the eye, and again after being reflected back.

The koala has a large, flattened, bare

nose. This is not only one of its most prominent features, but also one of its most important sense organs. Just as vocal communication is important to animals unable to see each other easily, so is communication via scent. For example, male koalas use their sense of smell to detect whether or not females are in their breeding cycle, and to recognise the presence of other active male koalas. But an equally important use of this sense is in the location and selection of food. Koalas carefully smell

The skull of the koala is large and oblong in shape and remarkably straight-sided. The gap (diastema) between the front and back teeth facilitates the movement of food around the mouth with the tongue. The large, flat molars are adapted for grinding, having crescent-shaped ridges on the surface as shown below.

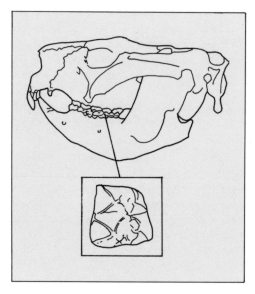

each leaf before deciding whether to eat it or not. This probably enables them not only to select the most palatable leaves but also prevents them ingesting some of the poisonous compounds concentrated in certain eucalypt leaves.

This is yet another area to be studied in depth, and until then we can only look at other marsupials, many of which have an extremely well-developed sense of smell. In some the olfactory bulb, which is that part of the brain given over to receiving and processing information from the nose, occupies almost half of the forebrain. It seems likely that the koala will be found to have a similarly well-developed scent-detecting apparatus, but this will only be known after further anatomical examination.

Dentition

The teeth of the koala are highly specialised for dealing with tough plant material. Like other herbivorous species they are more adapted to grinding than to tearing, and the sharp, pointed canines found in meat-eating mammals are of little use. On each jaw only one pair of incisors is well developed, the canines are very reduced, and two of the three pairs of pre-molars found in many other marsupials are absent altogether, leaving a large gap between the front and back teeth. This gap is called a "diastema", and is very important for an animal which has to deal with a tough, fibrous diet like *Eucalyptus* foliage. A diastema is also seen in other highly specialised plant-eaters, including sheep, cattle, horses and rabbits. It allows the tongue to move the long pieces of partially chewed plant tissue around in the mouth so that chewing is efficient; dealing with a large mouthful of such material without a diastema would not only be difficult and laborious, but the tongue would probably get bitten regularly.

The two pairs of rear teeth, or molars, on each jaw are large and oblong in shape, with crescent-shaped ridges on the upper surface covered with tough enamel. When the upper and lower molars are rotated over each other they grind the leaf material into a fine pulp. This arrangement of crescent-shaped ridges is termed "selenodont", meaning literally "moon-shaped tooth", from the Greek "selene" meaning "moon" and "odous" meaning "tooth", and is, like the diastema, typical of highly specialised herbivores.

Like most other marsupials, the koala has only one set of incisors,

In many ways the koala resembles the ground-dwelling marsupials, being short and stocky.
Yet it has long arms and legs, enabling it to climb high into the branches of gum trees.

The koala's feet and hands are endowed with long, sharp claws, without which it would be hard pressed to grip the often smooth tree trunks on its ascent into the arboreal world.

Koalas have some of the best insulating fur of all the marsupials. On cold days they huddle into a ball, exposing the darker fur of the back to the sun. On hot days they expose more of the white belly fur to reflect the heat, dangling their limbs loosely to increase the surface area exposed.

Koalas have a well developed sense of smell which is of great importance to nocturnal animals, enabling them to detect the presence of other animals in the dark forest, to recognise the scent markings of male koalas, and to locate and select the most palatable gum leaves.

canines and molars in its lifetime. One pair of premolars is, however, shed and replaced in the way that humans lose their "milk" teeth. The age of wild koalas has been estimated from their tooth wear by comparison with the tooth wear in known-aged animals, and interestingly, few males appear to live long enough for their teeth to wear out, although many females may live long enough to suffer from worn and inefficient teeth.

Food Selection

"If you awaken a koala towards evening, he may decide it is time he roused himself and got his breakfast. If he does, you will learn, probably for the first time, what fastidiousness really means. He gets out of his armchair and climbs upwards, slowly and carefully, clasping the boughs with arms and feet and sticking his strong, curved, needle-pointed claws into the bark. He stretches out a hand to a leaf, pulls it gently to that rediculous nose, and sniffs. No; too mature. He selects another; touches it with the tiniest tip of narrow pink tongue. No-o; a leetle too much body. He climbs higher. Tries again. Then again. At last! He munches solemnly, keeping even at meal-times his inimitable air of unfailing wonder, of innocent amazement, that any tree could be so crambed with surprises as the one he finds himself in at the moment."

The fastidious way koalas choose their food has rarely been described better than in the words of Dr Thomas Wood, quoted above. An Englishman, he visited Australia in 1931 and wrote an account of his travels in Australia in his book *Cobbers*. Just what it is that the koala is detecting with that "rediculous" nose is one of the most important questions that biologists must answer if the ecology of koalas, or any of the arboreal leaf-eating marsupials in Australian forests, is to be understood.

The extreme degree of selectivity displayed by koalas in their choice of food was obvious to the first naturalists to come across koalas, and has been commented upon innumerable times since. Koalas are generally considered to feed exclusively on *Eucalyptus* trees, but occasionally they have been reported to take the leaves of other trees including *Acacia costata*, *Bombax malabrica*, *Tristania conferta*, *Tristania sauveolens*, and even *Pinus radiata*, a pine tree introduced from North America for timber and which now forms forests that are unsuitable for most native mammals.

Koalas will only accept a small number of the hundreds of *Eucalyptus* species found in Australian forests, and of these only a handful will be accepted as the sole or major item in the diet. These staple species appear to be a necessary element of all koala habitats, and include the manna gum, *Eucalyptus viminalis* and the swamp gum *E. ovata* in Victoria and South Australia, and the grey gum *E. punctata* and river red gum *E. camaldulensis* in New South Wales and Queensland. It is impossible to present a more detailed list that would meet with general agreement among those studying the food choice of koalas because virtually everyone who has worked with or watched koalas for any length of time has a different list of *Eucalyptus* species they believe to be the koalas' favourites, and a different order of preference. The fact is that very few controlled investigations of koala food choice have been made, and those undertaken consistently indicate that it is not only the species of eucalypt that matters, but also the locality, the individual tree, and possibly even the site and the type of soil.

Koalas living in the Brisbane Ranges to the west of Melbourne have recently been studied, and provide a good example of the selective use of *Eucalyptus* trees by koalas. In this forest there were six species of eucalypt and a small number of other trees, such as *Acacia*. Koalas ate only eucalypts, but their preferences for species varied. The manna gum was most popular all year round except in summer when the swamp gum became most sought after. However, even when manna gum was preferred the koalas chose to live in areas with some swamp gum. Although these were the trends for the population in general, some individual koalas had quite different preferences. Certain individual trees within each species were favoured over others, and these tended to be the larger ones (possibly because they offered more shelter). Finally, the preferences of females changed during the breeding season, and those of males differed depending on their degree of dominance over other koalas.

The question of why koalas eat some species of *Eucalyptus* foliage and not others has attracted more speculation than any other aspect of the animal's biology, and is perhaps the most important question to answer if the koala's ecology is to be understood well enough for sites to be set aside and protected. Unfortunately this very question has also been the most difficult to address, and remains the least understood part of the koala's biology.

Many scientists now consider that plants are far from being passive victims of plant-eaters, and that they manufacture chemicals designed to defend themselves against being eaten. The type of defence would be expected to vary depending on the type of herbivore. A dose of an unpalatable or toxic compound may be enough to discourage a passing browser, but an animal like the koala that specialises in a particular group of plants takes much more discouraging. This may cost the plant dearly in terms of the potential nutrients that must be diverted to manufacture defence compounds, and in this case it may be more efficient to tolerate the animal but to minimise the amount that it can eat, while at the same time aiming to reduce its fitness, or the rate at which the animal can grow and reproduce.

In the light of these hypotheses it has been suggested that eucalypts manufacture toxic essential oils and some other highly toxic compounds to deter a range of insect and vertebrate herbivores, and that the tannins and lignin found in the leaves may function mainly to reduce the fitness of such specialist herbivores as Christmas beetles and koalas.

Many studies have looked into the koala's choice of eucalypt leaves to determine whether they prefer those with higher or lower amounts of such toxic chemicals. The results have, however, been contradictory and inconclusive, and it may be the case that animals are not as sophisticated in their choice of food as we presume. Research on sheep suggests that selection may be made on as simple a basis as how fast a food can be eaten. Thus, a characteristic such as leaf toughness may be much more important than any of the many chemical characteristics looked at.

Animals also choose trees for such reasons as shelter, protection and closeness to potential mates, and this may influence their food choice. Learning plays an important role in food choice in humans, and has recently been recognised to strongly influence diet selection in other animals. There may be at least a component of the koala's selectivity which is related not to the toxicity or nutrient content of the leaves, but to what it learned from its mother, or to its feeding experiences in the recent past. In this case, rejection of some species may merely reflect unfamiliarity.

Much more research is needed to unravel the mystery. But the possible approaches are far from exhausted, so there is hope that in the not too distant future management and conservation of koala populations based on more detailed knowledge of their dietary needs will be possible.

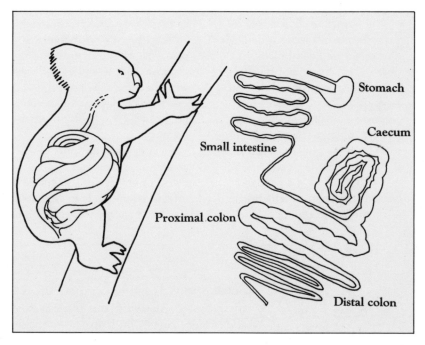

Digestion

The nutrition of any animal can be considered under three broad headings: the initial selection of the food, digestion of that food, and metabolism of the absorbed products of digestion.

Although the details of digestion and metabolism are complex, the basic processes are remarkably simple. The digestive tract may be considered as a system of tubes and chambers into which food passes and is mixed with enzymes, acids and other chemicals that break the food down into units small enough to pass through the walls of the tract and into the blood. From there these products of digestion are taken to various parts of the body where the multitude of chemical reactions of the animal's metabolism make use of the

The digestive tract of the koala is particularly well-developed to deal with its diet of eucalypt leaves. The caecum is among the largest of any animal, and plays an important part in digestion. Up to 2 m long, it occupies most of the available space in the abdominal cavity, and gives the koala its characteristic rotund appearance.

45

DEALING WITH TOXIC COMPOUNDS IN GUM LEAVES

Eucalypt leaves contain high proportions of two groups of chemicals that are quite poisonous, and unless inactivated may prove to be fatal. These are the essential oils and phenolic compounds.

The fatal dose of essential oil varies between species of animals. Humans have been killed by a dose equivalent to not much more than a teaspoon full, while over a period of 24 hours brush-tail possums can tolerate about ten times as much as this in relation to their body weight. But such is the concentration in the koala's diet that every day it probably exceeds the amount which would kill a human in one dose.

Among the chemicals known to be poisonous to humans, phenol is in the group with the highest toxicity rating. Ingestion results in nausea, headache, dizziness, muscular weakness, dimness of vision, ringing in the ears, irregular and rapid breathing, weak pulse, severe abdominal pain, loss of consciousness, collapse, and eventual death. As with essential oils, koalas probably exceed each day the intake of phenol which would kill other mammals.

Essential oils and phenolic compounds absorbed into the blood of koalas are recognised as foreign compounds in the same way as drugs are recognised in the blood of humans. The reaction of the body to such compounds is to attempt to render them non-toxic, and this is done in the liver by a group of enzymes called collectively the "mixed-function oxidase system". They remove or modify the part of the foreign compound which is responsible for its toxicity, and attach a small molecule to ensure that the new "conjugated" molecule is soluble in water and can be excreted in bile or urine.

If you observe areas where a koala has urinated, you will often notice that they are stained black. This is because large quantities of detoxified phenol have been excreted in the urine,

nutritionally desirable compounds, do their best to detoxify the undesirable and toxic ones, and excrete those compounds that cannot be beneficially metabolised.

The reactions are complex, but they amount to a chemical "burning" of fats, carbohydrates and protein to yield metabolic energy in much the same way that a fire releases heat energy to the air. Little wonder that an eminent physiologist once described energy metabolism as "the fire of life".

It has often been said that koalas eat several pounds of leaf each day. This simply is not true. The amount is more like 200-400 g (less than one pound) per day. They certainly remove a lot of leaves from branches, at least in captivity, but many of these are either accidentally dislodged or are taken in the mouth briefly and then rejected. Because of the relatively low energy content of Eucalyptus foliage koalas do have to eat more of it than they would

of a higher energy diet, but the koala's need for food energy is lower than those of most other mammals.

Digestion begins in the mouth as the leaf is chewed. The salivary glands secrete saliva containing enzymes to begin the break down of relatively simple compounds such as starch. Once the leaf has been chewed it passes into the throat and down a muscular tube called the "oesophagus". It next reaches the stomach, a relatively simple sack not much larger than a large hen's egg when full. Acids and enzymes are secreted into the stomach, and the breakdown of sugars and protein begins, that of starch continues, and some phenol and most of the essential oil from the leaf is released and absorbed.

From the stomach the food passes through a very long, narrow tube called the "ileum", or small intestine, where digestion of protein continues and the breakdown of fats occurs. This is also a major site for absorption into the

and the phenol has turned black on prolonged contact with air.

This process of detoxification places a heavy load on the liver, and it has been suggested that it may even cause gradual damage to this important organ. The liver of koalas is often dark purple in colour in contrast to the red of the liver in other animals, and this is probably a reflection of the intense metabolic activity taking place there.

But the possible effects of the detoxification process on the liver are only part of the problem faced by koalas. As if they do not have enough difficulty in obtaining energy, the process of detoxification uses some of it. The compound mostly used in the conjugation step is one called "glucuronic acid" and it requires

some of the koala's valuable glucose to synthesise it. Just how much energy this process requires has not been determined, but any drain is likely to be significant in the delicate energy economy of the koala.

It has not been determined whether koalas have a greater activity of the mixed-function oxidase system than other mammals which would enable them to deal with the large quantities of toxins they ingest every day. But it is known that the capacity of the system to deal with a particular foreign compound can be built up in all animals by regular exposure to that compound. A parallel can be drawn here with humans and alcohol. Regular exposure to alcohol increases the amount that

can be tolerated. Similarly, research on brush-tail possums shows that by sampling small amounts of *Eucalyptus* foliage each day they are able to gradually increase the amount of leaf they can eat before suffering ill effects.

The compulsive preference of koalas for a type of food containing so many drug-like substances has been likened to a long-term addiction. There was even a chapter in Ambrose Pratt's 1934 classic on koalas *The Call of the Koala* entitled "Koala's Addiction to Drugs". This notion is a bit extreme, but it probably is true that for koalas to maintain their ability to deal with large quantities of foreign compounds they must also maintain a regular intake of them.

bloodstream of the products of digestion.

From the small intestine the food moves into the hindgut. Here are found the most remarkable features of the koala's digestive tract and the keys to its survival on *Eucalyptus* foliage.

Firstly, there is the "colon", another long tube, somewhat wider than the small intestine, leading from the end of the small intestine to the "rectum". The beginning of this tube is enlarged to form a chamber resembling a large sausage. Off this chamber, at the point of its connection with the small intestine, runs another tube that ends blindly, and is the koala's claim to fame in the animal kingdom.

This is known as the "caecum". It is an amazing organ, being up to 200 cm long and 10 cm diameter. It is coiled up in the koala's abdominal cavity and, together with the expanded portion of the colon, occupies just about all of the space available. When the caecum is

full after a feed, the koala bulges like an inflated balloon.

The enormous size of the caecum was recognised by early anatomists, such as W.C. Mackenzie, who in 1918 described it as the most highly developed of all mammals. The naturalist Ambrose Pratt supposed that "a mysterious process of digestion must take place in this dilatation with its blind termination, to enable blood, bone, muscle etc., to be built up from such unlikely material." As might be expected, the Australian Aborigines did not overlook the enormity of the koala's caecum. In one myth the first Aborigines reached the east coast of Australia via a bridge constructed by a boy from the intestines of a koala. The crossing took several days, which means it was a very long bridge. After the crossing the bridge floated off into the sky to be the first rainbow seen by these people.

Today we know more about the processes occurring in the caecum of the

PRODUCING ENERGY BY FERMENTATION

The koala supports and nourishes a large population of micro-organisms in its hindgut. These microbes are essential to the koala's digestive process.

The process of digestion by micro-organisms in the digestive tract is known as "fermentation", and when it became known many years ago that fermentation occurs in the gut of the koala, many people thought they had found the reason for the sleepy disposition of the koala: fermentation, they understood, produces alcohol, and koalas are therefore perpetually intoxicated! Unfortunately, as appealing as this idea might be, it is simply not true.

The products of fermentation within the gut of the koala (as in those of other herbivores) bear no resemblance to alcohol, and are much more important to the koala's well-being. They are mainly short chain fatty acids which can be used to supply energy, and such gases as methane and carbon dioxide, produced by the breakdown of the otherwise indigestible fibre content of the eucalypt leaves. In humans this same process occurs, but on a much reduced scale. Herbivores, however, rely on fermentation to supply a large amount of their energy requirements.

Eucalypt fibre contains a high proportion of a substance known as "lignin". Lignin is particularly tough and indigestible, and comprises some ten percent of the fibre of most leafy vegetables, yet the fibre of *Eucalyptus* leaves contains more than 50 percent lignin. This puts koalas at a severe disadvantage compared to the other herbivores because the lignin binds to the other components in the fibre and prevents them being digested.

So despite the size of the koala's hindgut, the fermentation process only provides some ten percent of the energy they derive from eucalypt leaves. But with such a poor diet this may be an important element in the koala's survival.

koala, but there are still many mysteries. It is a region where digestion is carried out by micro-organisms rather than the koala's own enzymes.

Mammals have not evolved the enzymes necessary to digest the carbohydrate of dietary fibre (cellulose), but over millions of years have developed a mutually beneficial association with micro-organisms. The micro-organisms break down the carbohydrate and retain some of the products for themselves, but make the rest available to the animal in the form of compounds that may be absorbed from the gut and used for energy. Micro-organisms also manufacture other beneficial compounds, including some vitamins, that are used by the animal.

All plant-eating animals have a region housing a large population of micro-organisms. In animals like sheep, cattle and kangaroos, it is an expanded portion of the stomach, while in such animals as horses, rabbits, wombats, possums and the koala, the main microbial population is housed in the hindgut.

The large caecum of the koala probably has other functions, but research has yet to make these clear. For example, it has been suggested that this large chamber may be used as a water reservoir to be absorbed in times of drought when leaves are low in water.

The lower part of the colon extracts water from the gut contents, reducing the sloppy material found in the caecum to the hard, nut-like faeces that eventually emerge. This is vitally important to an animal with a limited water supply.

Metabolism and the Koala's Energy Requirements

Koalas seem to be able to survive quite comfortably on a diet that gives them only about half the food energy needed by other mammals. To exist on such a diet the koala has come up with a novel, but in fact absolutely logical answer: it has a remarkably low metabolic rate.

To produce energy by "burning" the fats, sugars and starch obtained by the digestive process, all animals consume oxygen, and one way of estimating the total energy used by an animal, and hence its metabolic rate, is to measure the amount of oxygen used. When oxygen consumption has been measured in koalas it has been found that, in relation to their body weight, they use less than half the amount consumed by most other mammals.

The low energy requirements of koalas are only partly due to their leading a quiet life, and are mainly the consequence of a metabolism that runs at a much lower rate than that of other mammals. The importance of this adaptation cannot be understated; it means that the koala has evolved yet another way of getting around the physiological constraints preventing most other small mammals utilising *Eucalyptus* leaves.

Life Cycle

Female koalas go through several oestrous cycles each year, and for a short period during the cycle (a few hours in captivity, and probably one or two days in the wild) they are receptive to males, in other words they come into "heat". Each cycle lasts for around 27-30 days, being similar in length to that of humans. Unlike humans, however, the breeding season is confined to the warmer months, and the young are generally born in the spring and summer.

In most marsupials studied in detail, ovulation (the release of the ovum from the ovary) occurs automatically one, two, or a few days after the oestrous cycle commences. Koalas may be different in that ovulation has been suggested to be induced by mating with the male. This requires further investigation, and its significance is not as yet

clear. Little work has been done on the hormonal control of ovulation or subsequent events in the reproductive cycle of koalas, so this area remains something of a mystery.

Pregnancy lasts from 34-36 days. Koalas usually have only one young at a time. Twins are extremely rare, and their survival even more so. A female, after all, has only two teats, and if she has just weaned a cub only the unused teat will be small enough for the mouth of the newborn. Twins were born in a fauna reserve in 1965, but by the age of 4.5 months there was absolutely no room for two in the pouch. One would certainly have died of exposure had it not been for the determined efforts of the staff in keeping it alive.

Many of the compounds in gum leaves are potentially toxic to mammals and are detoxified by koalas in the same way drugs are by humans. This and the fact that food is fermented in the gut of koalas led to the erroneous notion that they are perpetually intoxicated or high on drugs.

At birth a koala is only about 2 cm long and weighs less than one gram. Like all other newborn marsupials it is naked and blind, with an overdeveloped mouth, underdeveloped hindlegs, and forelegs only large enough to be used to climb up into the mother's backward-facing pouch, where it attaches to one of the two teats.

Just before birth the mother's pouch enlarges slightly, and as soon as the baby enters the sphincter muscle closes as tight as a drawstring. If no baby had been conceived the same thing still happens: the female has a pseudo-pregnancy, the pouch expands, and at the time the phantom baby should have been born the pouch closes tight for several days. This phenomenon is not, however, confined to koalas.

The development of the young koala in the pouch is very slow compared to other marsupials. There is some suggestion that this may be typical of mammals relying on low energy diets, but this has not been verified. The young koala remains fully in the pouch for six to seven months.

It is a miserable looking creature that first ventures out into the world. Its fur is pressed flat against its skinny body, and is covered with tiny black specks, that also line the pouch. They are its own excrements for, unique among marsupials, the mother koala does not clean her pouch. Once out, however, the infant must maintain elementary hygiene, for the droppings soon disappear and, concomitantly, the fur turns fluffy and displays a soft, brown tinge that gradually disappears. Many wild adults have brown blotches on their fur which may hark back to their pouch days, but in captivity they lose it.

At first the infant emerges for only a few minutes at a time, returning to the pouch to sleep and eat. After six weeks or so it is too big to return. Then it rides on its mother's back and sleeps curled up in her lap the way an adult does in a tree. Suckling takes place with the cub either sitting in its mother's lap or draped over her knee, and lasts for up to 20 minutes at a time. If her limbs are in the way when it wants to return to her lap it grasps her fur, puts its head in between the offending limbs, and moves its head from side to side until either the mother raises an arm or the infant has to give up. Exactly the same movements are made when it wants to return to the pouch or take milk. Although the infant's activities often cause great discomfort to the mother she is always extremely tolerant.

Both milk and leaves are included in the diet at this stage, and the young is not fully weaned until about one year old. This is quite a long period of maternal care compared with other mammals of comparable size, even marsupials. The infant begins to grow rapidly after leaving the pouch, and its first teeth emerge.

The mother's milk gradually increases in thickness and protein content as the young koala grows. It is initially high in carbohydrate and low in fat, but changes when the growth rate begins to increase to being high in fat and low in carbohydrate. Thus the very young koala is not oversupplied with energy, but when it emerges from the pouch the energy content of the milk increases to meet its requirements.

There is no evidence that any of the unusual and potentially toxic compounds contained in *Eucalyptus* foliage come through in the milk, but at the end of pouch life a very interesting event takes place which may well give the young something of a taste of the diet which is to come later.

Just prior to first eating leaves, when the young koala is about five to eight months old, its mother produces a special, very sloppy kind of faeces that the infant greedily devours, sticking its head out of the pouch and manipulating the dark green substance with its

The koala's head is large in comparison to the rest of the body, although it contains one of the smallest brains relative to its size of any mammal. There are very few whiskers, the eyes are cat-like, designed for night vision, the ears are large and the koala's hearing is acute.

Koalas are born in a very undeveloped state after a pregnancy lasting only five weeks. They are about 2 cm long and weigh less than one gram, yet they manage to crawl unaided from the cloaca to the pouch where they remain for six to seven months. The photograph shows a newly-emerged infant.

paw as it buries its face in it. This occurs over a period of 1-6 weeks, and the substance is thought to originate in the caecum and to contain the micro-organisms needed by the young koala to digest its diet of gum leaves. Even so, it still takes milk with its leaves for many more months. Alternatively, it has been suggested that it may play a part in the learning of food preferences by the young koala.

Female koalas, like humans, are usually unable to ovulate while producing milk. If she had a baby late the previous season she will thus not have time to wean it and have another before the next breeding season is over. She can therefore only breed two years running if the first young was born early or midway through the season, or if she is unlucky enough to lose the first one.

Young males, although they appear to reach sexual maturity in their second year, may not have access to females until they reach physical maturity at about four years of age. Females cease growing at the age of three. Their ovaries have been recorded as active as young as 13 months, although they first conceive at the age of two or three, depending on such things as health, population density, and whether she was born early or late in the year. In a thriving population the number of first time breeders is high, and it is not uncommon to find 60-80 percent of adult females with young every year.

The sex of the animal also determines its fate. Most areas have 10-20 percent more adult females than males. It is estimated that in a healthy popula-

tion 80 percent of the young survive to be weaned, and most of these survive until their second birthday. The maximum age for females is about 15 years, but males rarely live as long as this.

The Male Reproductive System

Very little study has been made of the reproductive system of the male koala, although it was described in detail as early as 1879 by A.H. Young, and several points are worthy of special comment.

The penis in male marsupials ranges from being an organ with a single, tapered head in some species to one with a forked (bifurcate) head in others.

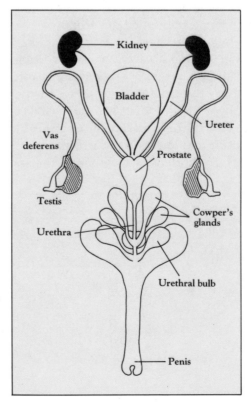

The urinary tract and male reproductive organs of the koala. The penis is forked at the tip, and the prostate is heart-shaped, unlike in most other marsupials where it is carrot-shaped.

THE PROBLEMS OF FIBRE AND SMALL BODY SIZE

Fibre occupies a considerable amount of space in the digestive tract, and even if it is digested it yields energy slowly. The total amount of energy an animal can extract from the fibre depends on how much food can be held and fermented in the gut at one time, or, alternatively, the rate at which the fibre can be excreted from the gut to make way for other, more digestible components.

The larger the capacity of the digestive tract the more food can be processed per day. Thus, animals that have evolved to feed on fibrous plant material have also evolved large digestive tracts. But there must be room for other organs in the body cavity, and so in no herbivorous mammal does the digestive tract plus contents exceed 20 percent of the body weight. In other words, small and large mammals are on roughly equal footing with respect to the amount of food they can process in relation to their total body weight.

Although small mammals may be able to process as much food in relation to their body size as large mammals, they need something like 20 times more energy per gram of body tissue to fuel their faster metabolic processes. A high fibre diet may therefore provide no problems for an animal the size of an elephant, but an animal the size of a mouse would be quite unable to supply its energy requirements with an equivalent diet.

Most small mammals live on insects, fruits, flowers and seeds, all of which contain little fibre and are high in available energy. Conversely, the animals grazing the fibre-rich grasslands of Africa or Australia are all large. The koala falls between these two extremes. It feeds on a high fibre, low energy diet, and therefore it should be large, but it lives in trees, and therefore it should be small and light enough to climb and move around without falling to the ground. But this same small body size means that it has a problem obtaining enough energy to survive.

The penis in the koala is intermediate between these two extremes.

The testes are located in an exterior scrotum, positioned in front of the penis, as in most marsupials, rather than behind it as in most placental mammals. As with other mammals in which the scrotum is exterior to the body cavity, the koala has the flexibility to vary the temperature of the testes by raising and lowering the scrotum.

This shortens or lengthens a remar-

kable network of fine arteries and veins which are intertwined between the body and the testes. This is the so-called "rete mirabile", and it is a counter current exchange system, meaning that heat in the warm arterial blood coming from deep within the body is transferred ferred to the cool blood returning from the testes, and this minimises the transfer of heat to the testes. The longer this network of arteries, the more efficiently heat is kept in the body and away from the testes. This amounts to air conditioning for a vital part of the body that will not function correctly at high temperatures.

The testes provide sperm, but the fluid making up the rest of the semen is contributed by the accessory glands: Cowper's glands and the prostate (other accessory glands are involved in the placental mammals). The prostate plays an important role in marsupials, and is correspondingly larger than in the placental mammals. The prostate

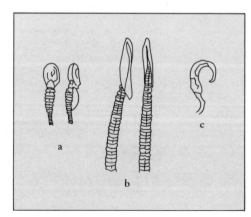

The sperm of the koala (c) differs from that of the other marsupials, being hook-shaped, and resembles that of the ground-dwelling wombat. Here it is compared with the ringtail possum (a) and the honey possum (b).

THE KOALA'S ANSWER TO THE PROBLEM

The koala has overcome the problem of small body size in several ways. It has a very reduced rate of metabolism, which means that it does not need as much energy as other comparably sized mammals; its sedentary lifestyle may be a means to conserve energy; and last, but certainly not least, the hindgut of the koala is able to separate fibre out from the rest of the diet and excrete it rapidly so as to provide space for more food.

The separation process seems to be accomplished by a series of contractions along the length of the hindgut causing waves to pass through the contents. Small particles are suspended in the current moving away from the rectum and into the caecum, while the large particles containing most of the fibre are moved towards the rectum. From there they can be excreted readily.

It is essential that the small particles escape from this rapid excretion mechanism because these include not only small pieces of cellular debris which can be digested by the micro-organisms of the hindgut, but also the micro-organisms themselves. If the micro-organisms were washed out they would have to be replaced by multiplication of those remaining, and this would mean that more protein would have to be supplied by the diet, and this in turn would compound the koala's problems.

The effectiveness of the mechanism can be gauged from experiments in which particles have been marked with radioactive markers. The fine particles of a given meal have been found to remain in the hindgut for as long as one month, whereas the large particles appear to pass out within 24-48 hours.

Koalas found dead in the wild often show signs of starvation even though the digestive tract is full of food. Cases like this may well be caused by the separation mechanism in the hindgut failing to work effectively, perhaps because of stress or disease, and hence the koala may be unable to maintain a high enough intake of food to meet its energy requirements.

grows in size during the mating season, and in the brushtail possum, for example, becomes the second largest organ in the body. It is likely that a similar enlargement occurs in koalas during the mating season. In most marsupials the prostate is a carrot-shaped organ, but in koalas and bandicoots it is characteristically heart-shaped. The significance of this difference, if there is one, is not known.

One aspect of the male reproductive system of the koala which distinguishes it from that in other marsupials is the sperm. They are unusual in shape in koalas, the head resembling a fishing hook, being similar to the sperm of the wombat, and this has lent more support to the idea that koalas and wombats are closely related.

BEHAVIOUR AND ECOLOGY

The koala possesses one of the most unusual behaviour repertoires of any mammal, but one that is entirely appropriate for its unusual way of life. An animal, for instance, does not need much intelligence or social organisation to sit around eating gum leaves, and the koala, predictably, appears to have little of either. Koalas sleep just as much by day as by night, but only a third of their feeding time takes place during daylight hours. The rest of the daylight hours are passed awake and resting. The average koala day, measured in the wild, consisted of 14.5 hours asleep, 4.8 hours awake resting, 4.7 hours feeding, and just four minutes travelling. Things get much more lively during the breeding season, but on the whole, koalas are, perhaps, the least active of all mammals.

A koala makes no nest. When resting it curls up like a ball in the fork of a tree, facing the trunk, its arms folded palms inward and the soles of the feet either facing outwards or pressed together. When the rain pours down it curls up tighter. When the sun is high it leans back, belly-up to the breeze, limbs dangling at all sorts of quaint angles, or even sprawls out with its stomach along the branch. But in all cases the weight of the body is taken on a tough patch of rump just above where the tail ought to be. Here the skin is attached direct to the bone with a tight layer of connective tissue, or gristle, and individuals have been observed sitting quite comfortably on the sawn-off tops of poles only a couple of centimetres across.

If an animal decides to stop on the ground, and this is very rare outside captivity, it normally sits like a dog, but may even curl up as if in a tree, or sit legs apart. On hot days some even stretch out on their bellies.

Individual feeding routines are highly erratic. Any time of the day will do, but during the hottest part of the day very little takes place. In some areas there is a peak of activity for a couple of hours after sunset; in others it occurs just before sunset and in the early morning.

Feeding is highly ritualized. First, a bunch of leaves is grasped with one hand (koalas are ambidextrous) and pulled to the mouth. The diner may then run its nose over the leaf and reject it if it is not to its fancy, but koalas are nowhere near as fastidious as popular legend makes out. The jaws then glide over the leaf and bite it off at the base, whence it is gradually worked lengthwise into the mouth by the rotary action of the jaws. Many of the older animals dribble, and their mouths and throats are stained yellow with *Eucalyptus* juice.

Baby koalas are a joy to behold. An infant's first contact with solid food occurs when some of the leaves its mother is consuming happen to fall within its reach. Without using its hands it takes a bite out of one, or else eats it without removing it from the stem. The next stage is when, by accident, it manages to hold down a leaf while still clutching its mother's fur. Later it starts to grasp the leaf beforehand, but not pull it to its mouth.

Whereas an adult always bites at the base of the leaf, a baby's bite is at first aimed haphazardly. It thus becomes difficult to sever the leaf, and so it has to jerk its head vigorously back and to the side, tearing large pieces out of the leaf. Because a leaf is so much larger for an infant it is also much more difficult to handle, and an infant can often be seen flailing its hands about trying to steady a leaf which should have been grasped firmly in the first place. By the end of the first year, however, it is feeding as competently as any adult.

Eucalypt leaves contain very little calcium, for which koalas have to descend to the ground. In captivity they have been observed lapping up gravel like water, and licking bricks and mortar, presumably for their calcium content.

I doubt if any living person has ever seen a wild koala drink, and it is popularly and erroneously believed that they never do. Koalas are superbly adapted to conserve body water. Most of their requirements come from the leaves themselves, and more can be licked from the leaves in the form of dew. This may, in fact, be the reason for scheduling their feeding times for early morning and just after sunset. Nevertheless, they do drink free-standing water, the tongue lapping at the rate of one or two strokes a second. And they do it for a considerable time: an average of four minutes, and a record of more than 14. Just the same, they can go without drinking for days in summer and indefinitely in winter.

On the ground they are flat-footed and slightly bow-legged, but they use the two basic gaits common to quadrupeds: walking by alternating a foreleg with the hindleg on the opposite side, and running by moving both front legs, then both back legs together.

When climbing the same gaits are used, but because they tend to extend the forelegs fully while keeping the hindlegs flexed, they often have to break step for the latter to catch up. Koalas will nearly always climb down a sloping branch on the upper surface, although they will sometimes climb up the same branch along the lower surface. Either way, they climb so their weight will be suspended from their

Resting postures of the koala. The variety of tree postures adopted depend on climate and conditions, (b) is that most frequently seen, while (d) exposes the maximum surface area for cooling on hot days. Different ground postures are shown in (e) to (h).

arms while the thrust comes from their hindlegs. For this reason they never come down head first.

Koalas have been observed jumping up to two metres between vertical branches, although this is uncommon. Normally they will either reach across to the next branch or stretch out and grasp the branch with their hands. Their agility cannot be doubted, and koalas have even on occasion been found at the very top of metal telegraph poles, a tribute to both their climbing ability and their single-mindedness.

Like most mammals koalas can swim using a sort of dog-paddle. This is only done in emergencies, however. Creeks and rivers are major barriers to dispersal.

Tame koalas will usually defaecate when they are picked up, a sign of mild nervous stress. Normally they only defaecate while resting, or even while asleep, and adults produce an average of 76 pellets a day. Urination, on the other hand, only occurs during activity. An animal typically halts in the middle of a climb, lowers its rump, and releases a copious stream of urine before moving on. Once a day is about normal.

Grooming is a pretty perfunctory affair. An inactive koala just lazily interrupts its rest and gives a few scratches to some part of its body, usually with a hindfoot, using the two grooming claws, or with a hand or the teeth. They never lick themselves nor, being solitary animals, do they groom each other.

Generally speaking, chewing and hand-scratching are used only for areas not readily accessible to the hindfoot. One of the more amusing experiments one can make is to gently pluck a koala's fur. Their aim is very haphazard. A foot will just rise blindly up and scratch vaguely in the general region of the irritation. If it is the flank which is irritated it may scratch literally anywhere, or even shake its head.

Koalas are not curious. Released into a strange area they will sniff the nearest tree without much eagerness to climb, and sniff the air at length, perhaps testing it for the odour of other koalas. They will prick up their ears and turn towards a strange noise, but not otherwise investigate it. During the breeding season individuals in captivity sometimes approach strange objects, such as shoes, and give them a thorough once-over of sniffing and chewing. As a general rule, however, strange objects, even placed directly in an animal's path, are totally ignored or, at best, granted a few perfunctory sniffs. Curiosity, in other words, will never kill the koala. But when you have few enemies and nothing better to do than eat gum leaves, why bother?

One of the more remarkable aspects of their behaviour is the ease with which they adapt to captivity. A wild koala is a dangerous beast to handle, but after a few months of regular, gentle handling will become, not friendly, but completely tame and placid, and may even sleep in a person's hand. Occasionally an individual will be seen pacing up and down in its pen, but there is none of the endless, neurotic pacing and obsessive, stereotyped activities displayed by many unfortunate zoo animals. Although basically solitary and antisocial, they can be raised in pens hundreds of times more crowded than the bush with minimal stress and aggression, even sleeping on top of one another. They are neither hostile nor friendly, merely indifferent.

Aggression and Display

We now have a picture of an animal that spends its time high up in a tree, well away from others of its kind, and which spends only a few minutes of each day doing anything other than eating, sleeping or resting. However, koalas do have a limited vocal repertoire, and communicate quite forcibly, particularly during the breeding season, and in captivity where the population density is far higher than in the wild it is even more evident.

It must be emphasized that nearly everything we know about their social organisation comes from the study of captive animals. To attempt to relate this to natural condition we have only guesswork and a few observations made incidentally to the investigation of other aspects of the koala's biology.

As with most other mammals, the breeding season is a time of major activity. Visitors to the bush at such times, particularly at night, may chance to hear an awe-inspiring bellow, something like a rhythmic belching and snoring sound, drifting down from the trees, and seldom think to connect it with the teddy-bear-like koala. If they are lucky they may be treated to a reply from some other section of the woods, for the sound is audible for at least 50 m, and sometimes 100 m on a still night.

Typically, a male wakes up, points his head to the sky, and inhales deeply, producing a long, tremulous "snoring" sound. Suddenly the air is expelled with a noise like a belch, his diaphragm contracts sharply, his head is jerked even further back, and another series of "snores" and "belches" follows. As the crescendo rises his nostrils flare, his voicebox heaves upwards, his cheeks are sucked inwards, and his tongue protrudes. The effect is thunderous.

A bellow, in other words, consists of a series of noisy inhalations and exhalations. Analysed on a machine called a sonagram, both parts will be found to consist of pulses so close together that they appear to be a continuous sound. They are formed by the rapid opening and closing of the vocal chords. A typical bellow lasts 10 to 20 seconds, with a record maximum of 2.5 minutes, and a male can give two or three bellows in succession, separated by just

Male koalas produce a thunderous bellow, particularly during the breeding season, with the head thrown back and diaphragm heaving.

a few seconds' pause for breath.

At the Lone Pine koala sanctuary bellows are heard only once every two hours for the first six months of the year, but by October the rate is up to four or nine per hour, depending on the year, and starts to tail off by December. Quite frequently the observer will be met with the curious and amusing sight of a young male putting all his heart and soul into a bellow, straining to the utmost, but all that comes out is a hollow, rattling noise like wind in a tube. Presumably male hormones are needed to motivate bellowing, but even more are required to enlarge the vocal organs sufficiently to make a proper sound.

Bellowing begins abruptly at the age of three and reaches a peak a year or two later, after which it gradually declines in frequency. Fully mature males at Lone Pine begin bellowing in earnest about July, but three-year-olds are still a bit immature then, and wait until September. One of them, however, kept it up till February, by which time he had the field very much to himself.

Most bellows are spontaneous, although they are more common on days of frequent bellowing, fights, or other vigorous activity. And in areas such as zoos where koalas are close together, males tend to reply in kind to one another's bellows, like a neighbourhood of barking dogs, until the whole place reverberates to the sound.

Aggression also elicits bellowing. When males fight generally only the victor bellows, but a male will bellow even if he loses a fight with a female, for those battles are less intense. Bellows are sometimes made at the mere sound of a nearby fight, or after an attempt at copulation.

The best hypothesis put forward so far is that the call is given when the bellowing drive reaches a certain threshold. After a few bellows the drive abruptly drops, then slowly rises again, but it will be several hours before the bellowing threshold is reached again. However, certain stimuli, such as aggression or the sound of another bellow, can boost the drive higher so that it reaches the threshold earlier. If the stimulus occurs when the drive is almost at threshold level a bellow will follow straight away. A comparison can be made with hunger, which drops to a minimum right after a meal, but the longer the time that elapses the more easily it is aroused by the presence of food.

Bellowing is neither a war cry nor a love song. It certainly acts as a challenge during a lull in a battle, and may arouse interest in a female on heat, but mostly the only effect is to elicit a response in kind. Its main purpose seems to be to indicate the presence of the bellower.

The bush is a lot quieter place than a koala zoo, and in any particular area only one male normally bellows: the mature resident. The juveniles and the nomads know that a worthy opponent is in the area and they keep quiet.

How did bellowing evolve? To answer this question we have to examine the behaviour of the females, for they have a much greater vocal repertoire than the males. Females bellow less than one-tenth as often as males, and usually with a much lower intensity. Although they are an integral part of sexual behaviour, the great majority take place during aggressive encounters.

Typically, as a female grows more angry her aggressive cries become louder. Then suddenly she takes a deep breath, automatically producing the "snore" part of the bellow, followed immediately by the "belch". Females often face their opponents while bellowing, and frequently there is an obvious conflict between the drives to stretch the head back and to face forward. In similar contexts sounds intermediate between a bellow and some other call have been recorded.

The koala's vocabulary has presumably passed through a long and complicated evolution, and males probably once bellowed in the same fashion as females. The female bellow is a sound induced by high excitement and anger, and the stretching back of the neck may be an automatic attempt to relieve the tension on the vocal organs and allow the free passage of air. Evolution could have acted to increase the volume of the sound, and in males the bellow has been freed from the aggressive drive, and is now done for its own sake.

Field workers are likely to hear cries not recorded in captivity. Koalas have been heard to make a sound similar to the chirping of a bird when trying to elude capture by humans. Individuals captured and released will often race up a tree and produce a series of high-pitched yelps interspersed with a chattering sound, often accompanied by shivering and shaking. This may continue for an hour and a half. Although this has been dubbed a "distress call", the function is completely unknown. If

Infant koalas emerge from the pouch at around six to seven months, for brief periods at first, but after a further six weeks they are too big to return, and ride around on their mother's back, sleeping curled up in her lap.

Koalas use the same gait when climbing and walking: alternating a foreleg with the hind-leg on the opposite side.

Despite their generally slow and sleepy disposition, koalas are, in fact, very agile, and have been observed jumping quite large distances between branches.

The male koala's scent gland is clearly visible as a large patch in the centre of the chest, and oozes a strongly-smelling oily liquid used to mark tree trunks in their home range.

SCENT-MARKING

Koalas have a very well developed sense of smell and a limited vocal repertoire, and like many mammals use scents produced by glands in the skin to communicate such information as their territorial boundaries in the case of males, and their receptivity to copulation in the case of females.

Male koalas have a scent gland in the skin of the chest, the "sternal" gland, with which they mark their territorial boundaries, particularly during the breeding season. The function of this behaviour is the same as that of a dog lifting its leg and urinating on various objects. It is the male's calling card, serving notice to other males that this is his stamping ground, and making the area familiar to him.

The gland is visible as a triangular patch of bare skin about 5 cm x 3 cm in the centre of the chest, which oozes a clear, oily, strongly smelling liquid. Secretions from the gland cause noticeable staining of the surrounding fur, allowing observers to easily identify male koalas.

Adult male koalas mark trees in their home range with secretions of their chest gland by grasping the tree and rubbing their chest up and down against the trunk.

Scent-marking is usually spontaneous, but often takes place in conjunction with bellowing, and after such stressful activities as fighting, sexual encounters, or handling by humans. The male approaches a tree from the ground, grasps the bole with his hands, flattens his chest against it, and rubs up and down as much as half a dozen times. Marking is automatic on wide-boled trees as the male cannot climb without pressing his chest against the trunk.

Recent studies at San Diego Zoo have shown that males are most likely to mark when introduced to a new environment, but their behaviour is not obviously affected by the presence of another male's odour. The experimental animals also tended to urinate copiously as they climbed the tree after marking. More rarely a male would simply touch his groin against the trunk and leave a spot of urine.

Scent-marking begins in males as young as 18 months, reaching a peak of activity at the age of four, corresponding to increases in such breeding season behaviour as bellowing and mating. In very rare instances females, most likely when on heat, have been known to go through the motions of scent-marking, although, of course, they have no sternal gland.

two adults of the same sex find them-selves sharing a tree they are likely to stop feeding and sit still while one or both makes an interrupted murmuring sound like the droning of an insect.

Soft grunts are made by all koalas with the mouth closed as a sign of mild displeasure. Very young koalas produce a sharp little squeaking noise lasting a tenth to a fifth of a second whenever they are slightly distressed. In older individuals this becomes a louder, harsher squawk, and indicates mild aggression. Adult males rarely make this sound.

Males do not make threatening vocalisations, although they produce harsh grunts during fights, a low sound given with the mouth closed.

Stand-off threats in the form of snarls, wails and screams are the exclusive domain of the females. A snarl is a rasping sound, best imitated by expelling air with the back of the tongue raised. A wail is a long, drawn-out, piercing cry consisting of several harmonics, and sometimes with a plaintive quality. It tends to merge into the scream, which is louder and even more prolonged. These calls are motivated by the conflicting drives inherent in all threats. Generally speaking, the scream is a sign of acute fear, the snarl denotes anger, and the wail a mixture of anger and fear.

Koalas do not have very expressive faces. When snarling they raise the corners of the upper lip, an expression seen to a lesser extent with all signs of mild aggression, such as squawks and some wails. A really furious individual, however, will have its lips pointed well forward, almost in the shape of an "O", and accompanied by bellowing or gurgling noises. The ears are also pointed forward, directly at the enemy. Conversely, when terror prevails, both the lips and the ears are pulled back. This habit of thrusting the movable parts of the face forward with anger and pulling them back with fear is very common among mammalian species, including our own.

Although koalas are generally very placid, fights do occur in captivity, and they are quite violent. In all cases the basic pattern is the same: an arm is thrown over the victim, which is then bitten on whatever part of its anatomy is closest. Everything else is simply a variation of force, orientation, and manoeuvring, and although fights may be violent they rarely result in serious injury. The koala's dense fur protects it against everything except the occasional scratched nose, swollen ear or, rarely, a piece of missing skin. Human beings, however, are not so well padded, and no untrained person should ever attempt to touch a wild koala. Their jaws are strong, and although the powerful, curved claws are never used in combat, they can cause severe damage if a frightened koala tries to climb on its captor.

Aggression is first seen at about the age of 11 months. In a crowded enclosure hundreds of minor squabbles occur when one individual tries to climb past another. They may last for several minutes, but more often than not the attacker is half-asleep at the time, and often the victim continues climbing without paying any attention at all. Generally speaking, a koala will only fight if it can see its opponent's face. An intruder, for example, may climb down a tree and push in between a sleeping companion and its branch, causing it a great deal of inconvenience, but because its back is turned towards the victim it will not be bitten. On the other hand, if an active koala finds its path blocked by a sleeping individual, it may sometimes give the unsuspecting sleeper a nip. A third party sitting next to a squabbling pair may also be bitten.

When fights break out between males they are either quite mild or else very violent, with few intermediates. The males wrap their arms around each other's shoulders, clinging tenaciously to the opponent's neck with their teeth. Harsh grunts issuing from their throats, they roll over on the ground, kicking at the other's abdomen with their hind feet in an attempt to regain footing. Sometimes they rear up on their hind legs, each pressing forward on the other while they grapple like wrestlers.

Then, unexpectedly, they may disengage, and one of them will vent his fury with a thunderous bellow, an action which usually brings his enemy back at him in a flash. Otherwise, a chase may ensue, with the chaser seeking to throw his arm around his opponent and sink in his teeth. Then, just as suddenly, the pursued may turn and become the pursuer.

Occasionally extremely violent fights break out among the males. Such battles normally last no more than ten minutes, and end when the loser breaks and flees or when the winner forces his opponent onto his back, biting savagely at his belly, whereupon the loser, unable to escape, screams at the top of his voice. The loser can then make his exit while the winner proclaims his triumph with a bellow.

Fights are, however, quite rare, even in captivity. Occasionally they are unexpected and without apparent provocation, being more likely to involve strangers newly introduced to a pen. One particular male at Lone Pine was moved to three different pens within the space of half an hour and became involved in a violent fight in every single one. The sound of fighting elsewhere tends to excite males, and a third party sometimes throws himself into the middle of a fight already in progress.

Females are more aggressive, but not as violent. Unlike males they use vocal threats and do not initiate fights. Threatened females often stand their ground, but will not move out of their way to attack. The retreat of an opponent is always sufficient to calm them.

Although females are normally placid and inoffensive creatures, at times, for reasons completely unknown, they become exceptionally aggressive. At such times they snarl, wail, scream, or even bellow whenever another individual approaches, or even simply moves around nearby. If the other koala continues its approach it will be attacked, often savagely. Such an attack will invariably force another female to retreat, but a male may fight back. Indeed, males have been known to make violent, completely unprovoked attacks upon females in captivity.

Fights between males and females in captivity are the result of the artificial conditions of such confinement. In the wild a female in her own tree would be better able to deter a male, for he would be out of his own home range and therefore at a psychological disadvantage. The innumerable minor squabbles are also an artifact of captivity. In the wild koalas would be much less familiar with one another, and their fights, though rarer, would be concomitantly more serious.

Mating Behaviour

Koalas are unique among mammals in that courtship is performed by the female rather than the male. In captivity the females are on heat (oestrus) for only a few hours, at which time, and no other, the female will copulate with a male. During oestrus females become very agitated and go through a repertoire of behaviour patterns that typically include bellowing, jerking, pseudomale behaviour and aggression.

The most unusual is jerking, in which the female holds onto the tree while her whole body jerks forward vigorously at the rate of about once a second, her head jerks back slightly and her ears flap. This behaviour has, in fact, been mistaken by visitors to koala parks as hiccups. Pseudomale behaviour consists of mounting, neck-biting and thrusting on another female, although usually briefly and incompletely. Such behaviour is not uncommon in other mammals such as cows. Mild aggression is displayed towards the male before copulation, and consists of chasing, throwing her arms around him and biting him on the neck.

Once a female is on heat this pattern of behaviour appears to be triggered by the presence of a male, the sound of a male bellowing, aggressive interactions or the appearance of strange individuals. It normally doesn't take long for a male to respond, and a healthy, fully adult female will conceive on the first coupling, although two-year-olds often fail. If so, they come into heat again at intervals of 26 to 30 days until the mating season is over.

Copulation is brief for marsupials, lasting only about 1.5 minutes, and takes place in a tree. The male mounts the female in a vertical position, clinging to the branch, and grasps the back of her neck in his jaws. The female then stretches her neck right back and raises her rump. The male makes vigorous pelvic thrusts about once a second, building up to about twice that rate. At each thrust the penis is inserted up to its base, then withdrawn half its length. As soon as thrusting is over and intromission complete, the female has a fit of convulsions. Her loins contract about once a second, and her head is jerked back even further each time.

Finally, the pair uncouple, although the female may have to squeal to make the male release his neck-grip. When she tries to leave the male generally puts an arm around her and gives her a bite, not necessarily hard, and a sudden fight breaks out before the couple separates. As with other marsupials, a plug of coagulated semen develops in the female's genital opening.

In captivity males frequently try to mate with females even though they are not on heat. There are no preliminaries and no courtship, just a resolute approach, mount, and series of pelvic thrusts. The female's resistance normally ceases once the male grasps her neck in his mouth, but because she refuses to raise her rump it is impossible for him to achieve intromission. These attempted copulations appear to be either spontaneous or precipitated by the added excitement of fighting or bellowing.

Males, even juveniles, frequently have penile erections, which may be quite spontaneous, but can also be elicited by fighting, handling, the sound of bellowing, and other forms of excitement. More rarely males may be seen jerking like an oestrous female.

All the behaviour patterns of breeding males, such as bellowing, scent-marking, penile erections and mating attempts, reach their peak in the fourth year of life, and can be elicited by the same stimuli.

The sexual life of the koala is thus very unusual. A neck-bite, for instance, normally only occurs in predators, which use it for hunting and carrying their young as well as for mating, and would never have been expected in the koala. Perhaps it enables the male to steady the female during convulsions. Those familiar with the behaviour of other mammals will not be surprised that the male's sex drive is intricately intertwined with aggression, and perhaps the female's behaviour is aimed at exciting him aggressively. Perhaps her jerking is a natural outlet for high excitement. Up in a tree, after all, it is impossible for the koala to run around or jump up and down.

The behaviour of koalas in the wild probably varies slightly from that described above. Oestrus almost certainly lasts longer in the wild, perhaps one or two days. Its very short duration in captivity is likely due to the ready accessibility of males, for a female normally loses interest once she has mated. The males' constant overtures to unwilling females probably occur only in the crowded confines of captivity where they are chronically overstimulated.

The story put forward in a number of popular books and articles that the males hold harems can certainly be discounted. Male home ranges do, however, overlap those of several females, and they also make exploratory sallies outside. Any female they encounter on heat will, no doubt, be given due attention. Even a young nomad male would have a chance to mate, and nomad females are known to have reproduced. Whether females on heat produce any odour for the male to notice, or whether they actively seek out bellowing males, only further studies will reveal.

Parental Behaviour

Young koalas do not appear to recognise their own mothers. Cubs in captivity will take milk from any female, and have been observed riding on the back of any adult. Females have been seen with two cubs on their backs, and all adults are extremely tolerant of the young. Mothers do not appear to recognise their own offspring, being prepared to accept any cub. This, no doubt, is a consequence of their solitary life in the wild where any young koala in the vicinity would most likely be the female's own. They do, however, have a "lost child response". In captivity mothers temporarily without a baby (which is most likely to be asleep on the back on some other female) can be seen wandering around uneasily, as if looking for something, and will hurry over to investigate any squeaks made by a lonely baby. If a child is placed on the female's back she will then usually settle down.

A charming myth of very long standing exists that koalas spank their young for misbehaviour. Ambrose Pratt popularised it in his well-read but grossly inaccurate book *The Call of the Koala*. Other stories include the throwing of the young from one parent to another, each parent in turn spanking the young koala! Needless to say, these reports are pure fantasy. Male koalas take no part in the rearing of their offspring, koalas never strike with their paws, and the concept of punishment cannot be applied to koalas. The only plausible explanation for Ambrose Pratt's spanking account is that someone observed a young koala draped over its mother's knee to get a drink of milk.

Play is a very important part in the life of many young mammals. It helps them learn social skills, investigate their environment, and develop skills in the manipulation of objects. Koalas, however, do not play with at all, which is perhaps not surprising considering their sedentary lifestyle, simple diet and lack of social organisation.

The bond between mother and off-spring is broken very gently. At about 10 months of age the infant begins making exploratory excursions away from its mother. These seldom last more than ten minutes or extend more than 50 cm. The infant moves from point to point, climbing, sniffing, and taking the occasional perfunctory bite from a leaf. It climbs on and off its mother's back and vacillates between her and the surroundings. If she starts to move it returns in a flash. Later the infant may become engrossed in feeding, and if it then stops and discovers that its mother has departed the panic and the squeaking begin in earnest.

When the mother's milk runs out cannot be pinpointed with accuracy. One young koala was observed suckling at 13 months, although this is probably very rare. Even after being weaned juveniles will climb on the back of any passing adult, which in the wild would usually be its mother. In captivity it is not uncommon to see a young, but very tolerant adult staggering under the weight of an overgrown juvenile. Some of these juveniles have bizarre habits. One of them, for example, gave up riding after being weaned but, still too young to breed, regained the habit in the next two breeding seasons, making a considerable nuisance of himself.

Distribution

The koala's ecological niche can be expressed very simply. It is one of only two species of mammals (the other being the greater glider, *Petauroides volans*) able to utilise the toxic, poor quality leaves of the eucalypts as its sole diet. It is bound to them like a sailor to his ship. Without them it has no existence. With them it has the freedom of the forest, with few enemies, fewer competitors, and even fewer needs.

Koalas do, however, need a certain amount of succulent foliage, and this is harder to find in the dry outback. In an area where the dew is light and free-standing water non-existent, they have to depend wholly on the leaves for their water supply, but the leaves are drier, and once eaten take longer to grow back. Even if the outback foliage is sufficient in a good season, the dry seasons are longer and the droughts more frequent, and in many places the trees are too sparse for breeding animals to find each other.

For this reason the wetter, seaward side of the Great Divide has always been the species' stronghold. To the west scattered populations exist, but the line of 50 cm annual rainfall appears, with few exceptions, to be the limit of their distribution.

When the whites first arrived koalas inhabited a belt stretching from Cooktown to the southeast of South Australia. They may never have existed in Tasmania and Western Australia, but the fossil record is too poor for us to be certain. With the possible exception of Stradbroke Island there do not appear to have been any populations on the offshore islands, though some islands have been artificially stocked recently.

The South Australian populations were shot out by the 1930s, leaving no clear records of their previous distribution. Koalas from Victoria were, however, introduced to Kangaroo Island at an early date, and are now doing fine. Further liberations have been made in several other places in the southeast.

In 1925 the population in mainland Victoria was estimated ("guessed" would have been a better word) at 500 to 1000, with only two colonies of any size. However, in the 1870s and 1880s private individuals artificially stocked Phillip Island in Westernport Bay, followed by French Island, and then the other islands of the Bay. These islands were the saviours of the Victorian race. Since the 1920s they have produced a surplus of more than 10,000

animals, to be liberated at nearly 70 different sites all over the State.

Today scattered populations can be found from the extreme east of Victoria to the extreme west. The heaviest concentrations occur in Gippsland, South Gippsland, and a broad area east and west of Melbourne, as well as in the central highlands up to Colac, Casterton and the Strathbogie Ranges. On Wilson's Promotory, where a single year once saw a harvest of 2000 skins, they no longer exist, and agriculture has virtually cleared them out of the Shepparton, Echua and Rochester districts. Nobody knows how many now flourish in Victoria, but the 2000 recorded in the 1967 schoolchildren's survey would be only a fraction.

In New South Wales the species is not doing too well. They have been shot out of the New England Tableland and most of the State south of Sydney, with the exception of a small area at the headwaters of the Shoalhaven. The vast majority occur east of the Great Divide from Sydney to the Queensland border, with further populations scattered in an arc from Sydney to Dubbo. Other populations cling to the banks of the Darling, Warrego and Culgoa Rivers.

The southeast corner is the stronghold of the Queensland race. In some areas, such as west of Rockhampton, it has suffered recent declines, but it is holding its own in the hinterlands of the Gold Coast and Mackay, and increasing in the Oakey, Springsure, Toowoomba and Maryborough districts. Scattered concentrations occur all along the coast up to Townsville and sections of the Atherton Tableland, and although the western regions have been greatly depleted some can still be found near Emerald and the Warrego River. For what it is worth, 2650 were counted in 1967 and 1123 in 1977. Both figures are presumably underestimates, but still somewhat less than the one million shot in 1919.

Present-day distribution of the koala includes a number of scattered populations, particularly in western Victoria and southeastern South Australia, reflecting the clearing of their habitat for agriculture and the reintroduction of koalas to specific areas.

The species is officially divided into three races: *Phascolarctos cinereus victor* in Victoria, *Phascolarctos cinereus cinereus* in New South Wales, and *Phascolarctos cinereus adustus* in Queensland. These were named by scientists with access to only a few specimens and with no idea of the extent of individual variation in any one area. Almost certainly they represent a single continuum from south to north, but the differences between the Victorian and Queensland races are striking. On the average the Victorian male is 70 percent and the female 90 percent heavier than their Queensland counterparts. The Victorian race also has a heavier, shaggier coat with more fur in the ears and around the face. Its face thus appears rounder and flatter. The Queensland koala is smaller, sleeker and much more attractive.

Population Structure

To study a population of koalas in the bush a zoologist has to develop the skill of sighting a dark ball of fur against the shadows of the foliage and the glare of the sky, and then attempt to get the koala out of the tree.

For this there are several accepted methods, all of which start off with a long ladder and an equally long, lightweight pole. If the end of the pole is furnished with a hook the animal can be shaken right off the branch and caught by assistants below with a fireman's net. The one and only time the author tried this the koala plunged at least ten metres, missed the net, and hit the ground like a block of rubber. Then, without being so much as winded, let alone injured, it was up and running away with the ground staff in hot pursuit. All of which shows that those little teddy bears are tougher than you think.

Nevertheless, broken bones have been recorded, and a safer, if more difficult method is to use a noose at the end of the pole. This is then slipped over the head of the koala, which is then gradually worked down the tree. An even more difficult method is to try to chase it down the tree by waving over its head a hessian bag, also attached to the pole.

Once the koala is on the ground the rest is easy. You simply grab it in such a way as to keep its claws and teeth away from your skin, weigh it, measure it, sex it, look at its teeth to estimate its age, take blood samples and do whatever else has to be done to obtain the desired information. Before releasing it you clip to its ear a couple of coloured tags so that it can be recognised next time without all that fuss. After that you have a new skill to learn: how to identify the colour of a tag at ranges of 10-20 m under the poor light conditions of the forest.

This, then, is one of the reasons so little is known about the ecology of the koala in the wild. But at least we know what density to expect. In the Oakey district of Queensland there is approximately one for each hectare of timber, much of it in strips along roads and creeks. On Phillip Island, Victoria, there are three per hectare, but this might have been a case of overpopulation. Certainly, at Walkerville, Victoria, a population of three per hectare crashed to less than a quarter of that level. In parts of South Australia a density of two per hectare can be maintained without damaging the environment.

Obviously one can't talk about density without considering the number of trees involved. In some parts of Kangaroo Island densities ranged from one animal per 79 manna gums to one per 17. Not surprisingly, the trees in the denser area looked well chewed around the edges.

Probably the densest population occurs at Tucki Tucki Reserve, 15 km south of Lismore. There 1000 trees, many of them very small, have been planted in just four hectares. It has been found that a stable community could exist with just three adult males and eight adult females in residence, along with their offspring and a few strays from outside: 27 in all.

It should not, however, be assumed that all the trees in a particular area are equally utilized. One tree may be eaten till it is in a pitiful condition while the one next to it, for all intents and purposes the same, is hardly touched. Taste is part of the the story. The more the tree is cropped the more juicy, tender shoots it puts forth. Provided they are not poisonous, why should the inhabitant move on?

In one zoo a female spent a whole twelve months within a metre of the same tree fork. Those who have to find their own leaves do a bit more travelling, but even in the wild they are loyal to particular trees. One famous study on Kangaroo Island found that many individuals were never seen away from their own special tree. Adult females had a home range, or area of normal activity, between seven and 176 m wide, but of all the trees in that area the female never occupied more then nine and sometimes only one. The width of male home ranges varied from 80 to 234 m, but only one to four trees were ever used.

These home ranges are not, however, distinct and exclusive, many trees being used by more than one animal. But being in the same tree at the same time is another matter. Koalas are solitary and antisocial. Most of the sightings made at Kangaroo Island were of solitary individuals, and most of the others entailed mother and young,

Catching a wild koala is no easy task, but is an essential part of research into the ecology of wild koala populations.

Koalas are very particular about the type and condition of the eucalypt leaves they will eat.
Often they will virtually strip one tree while hardly touching another of the same species
next to it.

QUESTIONS OF DOMINANCE AND TERRITORIALITY

Although the social behaviour of koalas is quite complex, their social organisation is relatively simple, revolving mainly around the protection of the young and the finding of mates. With few predators to guard against and no resources to find and defend, they do not possess the sort of community structure found in many other species, such as territoriality and dominance hierarchies.

Territorial animals have special areas or borders which they defend against other members of the same species. In a dominance hierarchy both dominant and subordinate animals live together, but by fighting and sparring they each learn their respective status. The dominants then have preferred access to food, shelter and females without having to fight.

There is no evidence that wild koalas have any borders to defend. In captivity they have no dominance hierarchy. Indeed, they lack the prerequisites: they have no appeasement gestures such as whining and cringing by which a subordinate can prevent attacks by a dominant, and the males do not have a repertoire of threatening gestures to deter other members of the group and hence prevent fights breaking out.

The koala, therefore, displays all the characteristics of a solitary rather than a social animal.

Male koalas do, however, have their own home range which they mark out with secretions from their sternal gland. Koalas wandering outside their home range have been observed to act nervously, and with good reason: the resident is likely to attack them. By means of scent-marks, bellows and general aggressiveness male residents produce a zone of intimidation around them which serves the same function as a territory or a dominance hierarchy.

How females maintain their separate home ranges remains a mystery. Their ranges may or may not overlap those of resident males, but they are certainly not under male protection. Females do not scent-mark, and if captive populations are any guide, they would not normally approach an intruder and attack it. However, their periodic and erratic phases of bad temper are defensive in nature and may perhaps keep them unmolested during the periods when their young are most vulnerable.

The young of the previous mating might then be eventually forced out of the mother's home range by her aggressive gestures and intimidation. If there are other aggressive strangers in the vicinity the juvenile will be forced to leave the area and adopt a nomadic existence until it finds an unoccupied area and establishes its own home range.

although the juveniles might have been nearly fully grown. Only seven out of 943 sightings involved two mature residents in the same tree.

In the rare event that two adults of the same sex find themselves in the same tree, one or both will eventually depart, although with breeding males the confrontation may well be violent. In the home range of a resident adult male, juveniles and strangers learn to keep a low profile. Even so, during population explosions, which are a regular feature of the Westernport islands, koalas can be seen as thick as fruit in the trees.

Home ranges and tree preferences are only part if the story. Even an animal as apathetic as the koala gets the wander lust occasionally. For one thing the quality of the browse changes throughout the year. In summer the Walkerville animals used to change from swamp gum to messmate and common peppermint, both of which were producing new growth. Similarly, the inhabitants of Kangaroo Island stick closest to their staple diet of manna gum in winter, and are more likely to venture into other areas during summer.

In all parts of Australia it has been found that koalas make excursions up to a kilometre or more out of their home ranges. On Kangaroo island individuals were seen to leave the

THE NOMAD KOALAS

Studies in all states have revealed that koala populations are divided into two classes: residents, with established home ranges, and nomads trying to find a place to settle. Males are more likely to become nomads, and when they do they take longer to settle down. Juveniles are the most nomadic, especially if they are male. The lucky ones manage to stay in their mother's home range, but most disperse about the age of two, and their fate is grim.

If the woods are full of resident adults, or if there are broad fields or stretches of unsuitable country to cross, nomadic juveniles can travel a long way; 48 km in 12 months is the known record. Many of these nomadic juveniles will die if there are no unoccupied areas with suitable food within a reasonable distance. All the dead koalas found in one study were young nomads.

In the Oakey region of Queensland half the males born on the site were still there at the age of two, one in eight were left at the age of three, and none at all by year four. They didn't come back. Perhaps some of them found homes elsewhere, because a number of strangers came from somewhere and took over their patrimony, but few of these stayed more than three years. The population was effectively divided between those under three and those over six, with few in between.

After a population crash the nomads may have an easier time, especially if they are the ones to have survived, or are immune to, the local diseases. Koala societies are therefore very fluid, with a constant turnover of population.

---

manna for the mallee for up to ten days. Females were off their home ranges more than a quarter of the time, males at least half. This custom of males having larger home ranges and being more mobile than females is very common among mammals.

Factors Limiting Population Growth

The oldest known koala was a female in San Diego Zoo that lived to the age of 17, and other zoos hold records not far short. But these are the centenarians of the koala world. The teeth of animals older than ten show severe wear, and malnutrition then becomes a threat to their longevity. Most zoo keepers would put the natural biological limit somewhere around 12 years. Determining the age of wild koalas is by no means easy, but one is known to have reached 13.

If a female first breeds at the age of two or three and has two young every three years, a fairly normal occurrence in healthy populations, she will produce six offspring by the age of ten. If two thirds survive to maturity, another reasonable assumption, and the offspring reproduce, the population could easily double in less than four years. In the 50 years since the cessation of shooting, koalas by this reasoning should have recolonised most of their old domain. Why, then, are there not now one or two koalas for every hectare of eucalypt forest on the east coast?

Firstly, only a very lucky koala lives to the age of ten, let alone 13. The majority perish in their third or fourth year while on the nomad trail, presumably because their homeless state makes them more prone to the natural hazards of life. However, this begs the question as to why there is nowhere to settle.

Habitat destruction is one limiting factor. Bush fires are always a hazard, and although burnt trees put out new foliage just as fast as koalas produce new young, a combination of disasters can be devastating. In 1970, for example, a cyclone hit Magnetic Island, and the accompanying rain stripped the trees of their leaves. The half-starved koalas were thus more vulnerable to a disease epidemic that followed. The dry weather, annual burn-outs and wild dogs also added to the toll so that between 1969 and 1977 the population fell from 1400 to 300.

Clearing of the land for agriculture makes some stretches of forest inaccessible. However, stands of timber along roads and watercourses act as effective corridors between useful habitats, a factor all farmers and local councils should remember. These stands are also the most vulnerable to *Eucalyptus* die-back disease. Even so, a lot of corridors exist, and vagrant animals have been known to cross fairly wide paddocks, especially if they contain isolated trees. Given enough population pressure, breeding nomads should be able to colonise a great many of the forested hills which abound in every rural district.

Koalas are not prone to much natural predation. Dogs are known to kill koalas, so presumably dingoes did the same at times. Eagles, hawks, and the occasional marsupial cat are capable of taking a young koala, but this must be very rare.

Aborigines used to eat koalas, and there is reason to believe that they once kept the population level down. They are said to have regarded it as a delicacy, although W.H. Brodribb, an early Victorian squatter, stated that their taste was far inferior to that of bandicoot or possum. The nineteenth century naturalist, Horace Wheelwright, claimed they tasted like the bears he had shot in Europe. (*Eucalyptus* flavoured bears?) Today the most dangerous "predator" is the motor car, but that only takes a few.

Disease has been noted as one of the major limiting factors in recent years, but our knowledge is so sparse that we do not know whether this is a naturally occurring level of infection or whether an epidemic is in progress. The cause of so much conjunctivitis and female reproductive failure may be due to one particular bacterium. The means by which the disease is spread is unknown, but it is reasonable to suppose that the ones most vulnerable are those suffering the stress of homelessness, the nomads, which are precisely the ones best suited to carry it from one area to another.

There are plenty of anecdotes concerning epidemics in the past, particularly in the years 1887-89 and 1900-03, and the descriptions sound suspiciously similar to present-day diseases. Disease, of course, is always the camp follower of malnutrition, and other anecdotes exist of koalas starving to death in places such as Quail Island and Wilson's Promotory after they had eaten themselves out of tree and home.

When a population is healthy its increase can be explosive, a good example being the Victorian islands, whence 8000 surplus koalas have been exported to the mainland. The trees on these islands did, however, tend to become defoliated. A well-watered eucalypt can replace its crown within three months, but the koalas won't let them. The four hectare Tucki Reserve managed to get its population up to 52 despite breeding an army of nomads for the inhospitable areas outside. Eventually 24 animals were moved to a new home, for the sake of the trees as much as for the koalas. Defoliation is also a problem in the South Australian colonies. Of course, the trees have to put up with leaf-eating insects as well, but the absence of reports of defoliation in former times is further evidence that the Aborigines kept the numbers down.

With increasing interest in koala research it was only a matter of time before one of these disasters took place in full view of a qualified scientist. That man was Roger Martin, who studies the population at Walkerville in southern Victoria.

These koalas managed to eat out and actually kill 40 percent of the local trees, especially the smaller ones. Compared to the well-fed population on French Island these animals were smaller in both weight and length, and the growth of their young was correspondingly retarded. Those who suffered most were the adult males which, being bigger, required more food. Between 1977 and 1980 they each lost between five and 12 percent of their weight. The largest animals were those in the stands of swamp gum, which had once been their staple food, but as time progressed they were forced more and more to messmate and common peppermint.

With starvation came disease. In August 1980 the females carried an average of 5.5 ticks each, and by January 1981 this had risen to 7.2. In contrast, a total of only two ticks were found on 71 French Island inhabitants. Much of the population was anaemic, but it was the sort of anaemia associated more with copper deficiency than with ticks. Martin then learnt that the messmate and peppermint they were being forced to consume contained only half the concentration of copper as swamp gum. Malnourished animals were more susceptible to disease, and at least 30 percent of the females were rendered sterile by ovarian cysts. In three years the population collapsed to less than a quarter of its original size.

Is there anything in the koala's social behaviour tending to limit population size? It has been noted that they tend to maintain a de facto "territoriality" in that the surplus population is forced into nomadism by the presence of aggressive adults. In other species territoriality function to defend some limited resource. The home range of any resident contains trees that are rarely, if ever, used. Younger animals could stay and eat them if they were not driven away, and they wouldn't have to die. One gets the impression that there is still a lot of unused space available.

How, then, can we explain the observations of dense populations of healthy individuals eating themselves out of their habitat? What about the evidence of captivity, where dozens live packed together in perfect harmony?

A reasonable hypothesis is that their social system not only affects the population density, but is itself affected by it. At moderate densities of, say two per hectare, koalas are intolerant. The home range in which they are prepared to harass strangers is more than sufficient to maintain them without permanent damage to the trees. However, in the wetter areas populations gradually increase. Each generation gets used to having its neighbours living just a little closer than before. Besides, the foliage is rich and green, so there is less need to wander abroad and have unpleasant confrontations. Eventually they become too tolerant, and the environment deteriorates. One might draw some comparisons with our own cities.

But the social system still exists, and the juveniles will still be pressured to disperse. At worst they will be driven to unsuitable habitats. At best they will stay in a good area but be driven from pillar to post. Neither is good for the health.

In a way the scattered nature of modern populations may be a good thing. Healthy populations can be kept isolated from sick ones. If enough research is done, and enough manpower expended, the surplus stock of the thriving areas can be harvested and introduced to uninhabited districts. The main thing is to find suitable areas and ensure that the number introduced is high enough to form a viable community. Australia's favourite marsupial is going through a bad stretch, but despite pessimistic reports it is not threatened with extinction, and with the help of mankind it may yet recover.

EVOLUTIONARY CONSIDERATIONS

Modern koalas may be somewhat mindless munchers of gum leaves, but they are also the last surviving branch of their own once enormously diverse family tree, the Phascolarctidae. In fact, they represent the only living descendants of an entire marsupial suborder, the Phascolarctomorpha (meaning animals shaped like koalas). To understand what this means, it might help to visualise a huge gum tree from which all but a single branch has fallen. This lonely remnant of a once full crown is more or less analogous to the living koala: it is the last branch left on what was once a very complex tree.

Initially thought to be some kind of monkey, koala affinities have since been forged with cuscuses, ringtail possums, the greater glider and wombats. Contemporary studies (including studies of sperm morphology, blood proteins, dentition and anatomy), however, have produced what now appears to be a relatively clear and unambiguous understanding of koala relationships: of all groups of *living* marsupials, the koala seems to be most closely related to wombats.

But, if we consider also the *extinct* groups of marsupials, the picture becomes a little less clear. The diversity of koalas as well as primitive members of other marsupial groups inclines us to be far more cautious about interpretations of the precise relationship between koalas and wombats.

The reason for this is that there appear to be some extinct groups of diprotodont marsupials (those which have two prominent lower incisor teeth

Current understanding about the evolutionary position of the koala among the marsupials is summarised in this diagram. There are at least two major groups of diprotodont marsupials: the wombat-like (vombatiform) marsupials and the cuscus-like (phalangeriform) marsupials. Although koalas are usually placed with the vombatiform marsupials, there appear to be some extinct marsupial groups with which they share a closer relationship, suggesting that they should perhaps form a third, independent group of diprotodont marsupials, as indicated here.

EXTINCT KOALAS, THEIR AGES AND THE PLACES WHERE THEY WERE FOUND.		
Species	**Age**	**Location**
Perikoala palankarinnica	15 m. yrs	Lake Palankarinna, S.A.
The Robust *Perikoala*	15 m. yrs	Lake Palankarinna, S.A.
De Vis' Mada Koala	15 m. yrs	Lake Pinpa, S.A.
Wells' Mada Koala	15 m. yrs	Lake Namba, S.A.
Mada Koala "Form C"	15 m. yrs	Lake Palankarinna, S.A.
Litokoala kutjamarpensis	14 m. yrs	Lake Ngapakaldi, S.A.
The Kanunka *Litokoala*	?14 m. yrs	Lake Kanunka, S.A.
The Riversleigh *Litokoala*	?14 m. yrs	Riversleigh, Qld
Koobor jimbarratti	4.5 m. yrs	Allingham Ck, Qld
Koobor notabilis	4.5 m. yrs	Chinchilla, Qld
The Sunlands *Phascolarctos*	?3-4 m. yrs	Sunlands, S.A.
Phascolarctos stirtoni	?50,000 yrs	Cement Mills, Qld

ANCESTORS OF THE MODERN KOALA

Perikoala palankarinnica
Stirton, 1957

Age: Approx. 15 million years
Local fauna: Ditjimanka Local Fauna
Geological formation: Etadunna Formation
Type locality: SAM Quarry North, Lake Palankarinna, South Australia

The first discovery of a fossil koala in Australia occurred as part of a most exciting field trip in 1953. Apart from the discovery in Tasmania of an early Miocene marsupial (about 23 million years old) and a few bits of marsupials from Beaumaris in Victoria (about 5-10 million years old), prior to 1953 there were no significant fossil land mammals known that were older than about 5 million years.

In that year, Professor Rueben Stirton from the University of California at Berkeley, accompanied by Dr Richard Tedford (then of the same institution), Geoff Woodard and Paul Lawson (of the South Australian Museum), struck it rich.

On 27 July, after weeks of fruitless searching, Geoff Woodard discovered a locality on the west side of Lake Palankarinna in the Tirari Desert of South Australia, that was loaded with the teeth and jaws of late Tertiary mammals. This was an exciting and long-awaited beginning to what was to become decades of subsequent research in the same region.

Tedford described Woodard's discovery in his 1953 field notes as follows:

"We were rather disappointed that these deposits had not yielded mammal remains and were ready to give the show away. We pulled up across a green clay bench where Woodard was prospecting to be greeted by a 'top of the morning' doff of the hat, big smile and outstretched hand bearing fragments of kangaroo bones and teeth. Geoff had done it and we had rolled over his hunting ground, a bench strewn with mammal bones, turtle shell and crocodile teeth. This was it, the first Tertiary mammals from Australia!"

As exciting as the moment was, at this particular spot, which appropriately came to be known as Woodard Locality, very few *different* kinds of mammals were found; and although they were indeed Tertiary in age (2-65 million years old), it turned out that they were relatively young, probably no older than 4 million years. But, immediately following this initial discovery, other fossils representing quite different kinds of mammals, including the first and still oldest-known koalas, were collected from different localities along the edge of Lake Palankarinna.

At first these other kinds were thought to be the same approximate age as the Woodard Locality fossils, although it is now known that they are about 15 million years old. Of particular interest here was the discovery by Richard Tedford of part of an upper jaw of a koala from a Lake Palankarinna site that came to be known as Tedford Locality. This specimen of *Perikoala palankarinnica* was not described until 1957.

A better specimen of the same species was discovered by Paul

that project forward from the lower jaw) that bear a closer relationship to the wombats. These include the group of extinct marsupials (the diprotodon-toids) to which the browsing, cow-sized *Diprotodon* belongs.

In fact, it is now becoming clear that there are at least two major groups of living and extinct marsupials: the vombatiform order (those that are more-or-less wombat-shaped and whose only surviving representatives are the wombats) and the phalangeriform order (those more closely resembling possums). All of the phalangeriform mar-supial groups look possum-like except the kangaroos.

While conventional wisdom suggests that koalas belong with the vombatiform marsupials, there is also some justification for suggesting that perhaps koalas should form a third independent order of diprotodont marsupials. While this is possible, until we know more about the record of fossil koalas prior to 15 million years ago, their precise relationship to the vombatiform marsupials may remain a mystery.

So far 12 extinct species of koalas have been discovered, spread over four

Lawson in 1954 at yet another locality on Lake Palankarinna which came to be known as SAM Quarry North Locality.

Paul recalls standing in the SAM Quarry North Locality in the summer of 1954 and swinging his pick into the sandy fossiliferous clays, to discover a large portion of a left lower jaw of a fossil koala. This became the holotype (name-bearing specimen) of *Perikoala palankarinnica* and is now in the palaeontological collections of the South Australian Museum. It was better preserved than the upper jaw fragment found the year before by Tedford and so was more appropriate as the holotype for the species.

In the years since these first discoveries, much work has been carried out at Lake Palankarinna in a continuing effort to recover more of the known species as well as additional new kinds. This extra effort has resulted in more koala material and *Perikoala palankarinnica* is now known from most of its upper and lower teeth.

A third substantial specimen, a lower jaw with three teeth in it, was found by Steven Van Dyck from the Queensland Museum in a most unorthodox manner. It was during 1977, on his first visit to the by then famous Lake Palankarinna fossil localities. Archer had just finished explaining to him about the extraordinary rarity of mammal jaws in the Lake Palankarinna sediments. In fact, it is extremely rare to find anything other then a few isolated teeth despite what may be weeks of hard work by experienced collectors. But when Van Dyck first set foot outside of the Toyota, it was to step straight onto the second known and best-preserved lower jaw of *Perikoala palankarinnica*. Such delightfully improbable coincidences are the stuff of palaeontological lore.

The jaw survived the encounter beautifully and is now a very significant addition to our knowledge of this otherwise poorly-known species.

The koala and its world
Although *Perikoala palankarinnica* is now known from almost complete upper and lower dentitions, this is still very little on which to base a reconstruction of the whole animal. At the time of writing, a serious scientific reconstruction of this koala has not to our knowledge been attempted. But, judging from its teeth and the other animals with which it is fossilised, it is possible to come up with what is probably a reasonable interpretation of the animal as well as the world in which it lived.

Clearly it was a browser, as were all koalas, probably feeding on the leaves of trees or shrubs. This browsing habit is evident from its selenodont teeth featuring very distinctive W-shaped cutting crests on the cheekteeth, specialised for leaf-eating. It was only about 80 percent as large as the living koala, to judge by the length of the tooth rows, but had a proportionately longer and more powerful premolar (the cutting tooth in front of the molars). This suggests the possibility that it may have had to cut more twigs or woody stems than the modern koala in order to get the leaves it required.

As for the rest of its anatomy, this will remain a mystery until significantly better fossil material is discovered. That this is possible has recently been demonstrated

genera, and dating back some 15 million years. Although there is not a great deal of material to go on, things are improving. About a dozen beautifully preserved specimens of the Riversleigh koala from northwestern Queensland are now under study, and this will at least markedly improve our understanding about the otherwise poorly known koala genus *Litokoala* (some 14 million years old). Unfortunately, it may take many years yet before more is known about the various kinds of koalas from the central Australian Miocene deposits. That material comes

with much less ease and is rarely well-preserved.

What can fossil bones and teeth tell us about the extinct koalas that once bedecked ancestral forests? Fortunately for those interested in the history of Australia's biota, quite a lot.

For example, differences in positions of muscle scars on bones tell us how the musculature, movements and sometimes behaviour of extinct forms differed from those of living species. Differences in the width and length of the space between the tooth rows indicate the shape and probable func-

by Neville Pledge's discovery of two almost complete skeletons of wynyardiids (dog-sized herbivorous marsupials) at Lake Palankarinna. If a similarly preserved skeleton of *Perikoala palankarinnica* could be found, much of the present mystery surrounding this the earliest-known koala would fall away.

Perikoala palankarinnica had many ecological associates that shared its space and time. All of these are part of what is called the Ditjimanka Local Fauna, an assemblage of creatures represented in the same stratigraphic levels at Lake Palankarinna. These include: a primitive platypus; primitive dasyuroids (insect-eating carnivorous marsupials); a miralinid (an extinct group of very strange possums distantly related to cuscuses); a primitive ringtail possum; an ektopodontid (extinct rodent-like possums); a pilkipildrid (extinct possums distantly related to the living gliders and ringtails); an ilariid (giant extinct diprotodontian marsupials that were distantly related to wombats and koalas); and even Australia's oldest bat.

Other kinds of Ditjimanka

mammals have yet to be named, including a kangaroo or two and a diprotodontid (extinct sheep-sized browsing marsupial). In addition to the mammals there were many kinds of fish, frogs and reptiles as well as wading birds.

In addition, analysis of pollen types found in the lower part of the Etadunna Formation (below the level at which the Ditjimanka Local Fauna occurs) suggests that there was a mixture of closed and open forest plants in the region. There is even some indication that grass pollens were in the area, though this does not necessarily indicate that grasses were present in the immediate area of Lake Palankarinna because, as wind-dispersed pollens, they could have been blown in from some distance away.

If we place all of the diverse evidence on the table, it is apparent that the Miocene countryside around what is now Lake Palankarinna was a far cry from the desert conditions that dominate the area today.

The landscape would have been partly covered by a large fresh to brackish lake (as

evidenced by the wading birds, fish, turtles, crocodiles and platypus). The lake apparently did not open to the ocean (as did lakes of similar ages around what is now Lake Frome) because there is no sign of marine organisms of any kind.

The lake would have been surrounded by vegetation including trees (as evidenced by southern beech (*Nothofagus*) pollen as well as ringtail possums) and shrubs (browse for the ilariids and diprotodontids). Insects were probably abundant (as food for the frogs, small lizards, small carnivorous dasyurids and bats). If there were songbirds in the area, they left no trace in the fossil record, so the "morning chorus" probably would have been largely grunts and groans.

There is no evidence that grasses were in the immediate area (there are no mammals here with teeth capable of eating grasses) so the ground cover may have been mosses or ferns. Rainfall was probably substantial and regular (as evidenced by the *Nothofagus* pollens), although possibly seasonal. It is possible that further away from the immediate site of accumulation

Four-wheel-drive tracks criss-cross desert sands at the base of eroded hills on the western shores of Lake Palankarinna, South Australia. From this area came the specimen of *Perikoala palankarinnica* discovered by Steven Van Dyke in 1977, a lower jaw of this 15 million year old koala which he stepped on when he first set foot outside his vehicle.

The first isolated teeth of the 15 million year old Perikoala robusta *were found in the lighter-coloured sediments known as Croc Pot 8, an area on the shores of Lake Palankarinna named after the abundance of crocodile teeth and skull bones found there.*

An RAAF helicopter lifts out just less than one tonne of fossil-bearing limestone collected by volunteers from an inaccessible plateau on Riversleigh Station in northwestern Queensland. Some 200 previously unknown species have been discovered in this area, ranging from 50,000 to 15 million years old. The Riversleigh koala, now known from a number of different sites in the area, may represent an evolutionary link between some of the Miocene koalas and the living species.

Along the quiet reaches of the Condamine River near Chinchilla in southeastern Queensland, sediments rich in 4 million year old mammal bones have revealed the remains of a very strange koala, Koobor notabilis, thought to have been collected in 1889, and initially described as a ringtail possum.

tion of the tongue. Size of and position of eye sockets tell us about the presence or absence of stereoscopic vision, night vision and other attributes that in turn tell us more about the probable behaviour of the extinct species. Examination of the size and distribution of channels in the bones for blood vessels and nerves can alert us to the presence of unusual structures like trunks. Study of the braincase can provide masses of information about behaviour, including whether the animal was an agile climber, a rapid runner or a swimmer. Study of the shape of the bones of the middle and inner ear can even tell us about the sounds that the ancestral creature was accustomed to receiving as well as about the probable position and size of the external ears. Analyses of the chemistry of fossil bones and teeth are even enabling palaeontologists to determine where in the prehistoric food chain the extinct animal belonged and, in some cases, on which types of vegetation a herbivore specialised.

In the case of fossil koalas, although little other than teeth and jaws have so far been found, using the modern koala as a guide, we can interpret the differ-

there were uplands that had grasses and grazing mammals. In which case, our picture of the area may apply only to a narrow zone around the lake margins.

In this sort of habitat it is likely that *Perikoala palankarinnica* spent its time eating tree leaves and gazing serenely out over the vast flamingo-dotted lake.

From life to fossil Fortunately, at least three individuals of *Perikoala palankarinnica* were so intent on gazing at this tranquil Miocene scene that they lost their balance and fell into the lake to become, 15 million years later, part of the Ditjimanka fossil assemblage.

Very few of the Ditjimanka fossils are intact. Most consist of isolated teeth. This, combined with the fact that the bones seem to occur in layers of similarly disarticulated but otherwise undamaged fish bones, suggests that the carcasses of the koalas as well as the fish accumulated on what was then a quiet bottom of the lake without interference from predators or large scavengers. Perhaps small animals like yabbies then cleaned the skeletons, allowing the undamaged bones to simply fall apart through natural decomposition. Possibly the decomposition of organic remains made the bottom of the lake unpleasant for large scavengers and hence a quiet final resting place for the remains of *Perikoala palankarinnica*.

The Robust *Perikoala*
Woodburne et al., 1987

Age: Approx. 15 million years
Local fauna: South Palankarinna Local Fauna
Geological formation: Etadunna Formation
Type locality: Turtle Quarry, Lake Palankarinna, South Australia

The first discovery of material referable to this species was found by Michael Archer, Michael Plane and Steven Van Dyck in 1977 when collecting at what became known as the Croc Pot 8 Locality. This locality was so-named because of abundant teeth and skull bones of fossil crocodiles.

The koala remains from Croc Pot 8 Locality are far from spectacular, consisting as they do of just isolated teeth. Although it has since been determined that they represent a species distinct from *Perikoala palankarinnica*, the species level distinctions were not evident at the time of discovery and hence the teeth remained unnoticed in a museum drawer, until the holotype was discovered.

The holotype of the Robust *Perikoala* was collected by Dr Michael Woodburne, Dr Judd Case, Mark Springer and other members of their 1985 expedition to Lake Palankarinna. It is a left lower jaw that retains most of the cheekteeth.

The koala and its world Very little can be said with confidence about the actual appearance of the Robust *Perikoala*. From the scientific description of the material, it is evident that it was most similar to *Perikoala palankarinnica*. However, judging from its teeth, it appears to have been a somewhat more robust species with wider and more elaborate molars, although the length of its toothrow is not

ences in shapes of teeth in the prehistoric koalas as indications of different dietary adaptations. For example, if the teeth of one extinct koala are lower-crowned with thin-enamel, we can suggest with reasonable confidence that it was not feeding on foods that were as abrasive as those utilised by the living koala. Similarly, to the extent that an ancestral koala's teeth display the W-shaped selenodont crown pattern, we can safely assume that it fed on the leaves of trees rather than those of grasses. Wider and shorter toothrows suggest a more powerful bite (bio-mechanics enable determination of the forces involved in biting and chewing) which in turn suggests that such species may have consumed leaves from trees that had branchlets more resistant to cutting than those of modern eucalypts.

This sort of data gathering is rather like detective work. First we must know the relationships between particular functions and anatomical specialisations in living animals. Then we must study and document all of the ways in which the extinct animal's bones and teeth differ from those of its nearest

significantly different from that of *Perikoala palankarinnica*.

Like other koalas, the Robust *Perikoala* would have been a browser on leaves, probably tree leaves, because we do not know for certain if *Eucalyptus* trees were present at this time, it is not even possible to speculate about the kind of trees that provided the Robust *Perikoala* with its belly-full of leaves.

The time in which the Robust *Perikoala* lived was prior to that in which *Perikoala palankarinnica* lived. This is evident from the fact that its remains come from sediments about three metres below those that contain the remains of *Perikoala palankarinnica*.

How this difference in time may have been reflected in the environment of the Robust *Perikoala* is uncertain. It is not yet clear that the animals present with this species were significantly different from those that accompanied *Perikoala palankarinnica*. Much more work needs to be done comparing the ancient ecosystems that replaced one another in the Lake Palankarinna region.

From life to fossil The situation in which the remains of the Robust *Perikoala* have been found at the Croc Pot 8 Locality is certainly different to that for *Perikoala palankarinnica*. At Croc Pot 8 the bones of all the animals now treated as part of the Lake Palankarinna South Local Fauna occur in discrete round to oval patches. This suggests a number of different scenarios.

Perhaps the bone accumulations represent the excreta of a meat-eating aquatic animal. After snatching animals that had fallen into the edge of the water, or scavenging on already dead carrion, the unknown eater of beasts leisurely digested its meal and then excreted the indigestible remains (the bones and teeth) as discrete bundles. These were subsequently buried and fossilised. This phantom carnivore, if it was responsible, was more likely to have been a turtle or lungfish than a crocodile, because modern crocodiles at least can digest bone. Alternatively, perhaps the discrete patches of bone represent depressions in the lake floor where bones settled after being moved around by currents.

Litokoala Kutjamarpensis
Stirton et al., 1967

Age: Approx. 14 million years
Local fauna: Kutjamarpu Local Fauna
Geological formation: Wipijiri Formation
Type locality: Leaf Locality, Lake Ngapakaldi, South Australia

The discovery of the Leaf Locality is a good example of how palaeontological advances are often made. One of the early discoveries of Stirton and his colleagues was that Lake Ngapakaldi, an otherwise inhospitable salt lake nestled among sand dunes in the Tirari Desert of South Australia, had a hard green claystone exposed on its eastern edge that produced fossil bones. Prior to this discovery, most of the central Australian discoveries had been made near the cliffs on the western sides of the salt lakes. Ngapakaldi Quarry, as it became known, was found on the east side of the Lake.

Once there and working, Stirton's colleagues had time to look more closely than they otherwise might at the rest of the flat-lying sediments just surfacing

living relative. The combination of all of this information enables us to develop an overall picture of the ways in which the extinct form is similar to or differs from the living relative.

As a picture of the extinct animal develops, it represents a testable hypothesis. Each new fossil found, or each new analysis of the bits already known, serves as a test of current understanding. When the test is over, the old interpretation may require significant modification, or it may have received a pat on the back. In this way, palaeontological interpretations themselves evolve through time, each change resulting from the never-ending process of data gathering and hypothesis testing.

In reviewing the extinct species we have tried to present not only a rough idea about what each of the various koalas would have been like, but also of what their habitats were evidently like. We should keep in mind, however, that because the study of these fossil koalas and their ancient environments is only just beginning, we must look forward to frequent and often significant changes in the interpretations provided

---

on the east side of the Lake. And it was not long before Paul Lawson found some bones and pebbles lying on the surface at a locality that later became known as the Leaf Locality.

When the bone-bearing sediment was excavated, it turned out to be a relatively rare type of deposit for the area: an ancient stream bed conglomerate topped with a fossil leaf-bearing shale. The stream deposit proved to be very rich in the remains of small mammals of a kind previously unknown in the other Tirari fossil deposits and, among the most interesting, was a single upper molar of a very distinctive koala.

This is a common way in which new fossil discoveries area made. A single discovery in a previously unknown fossil field encourages closer examination of the whole area and, as a result, yet more discoveries are made in areas that might otherwise have been overlooked.

There were several returns to the Leaf Locality with the resultant discovery of many more new kinds of middle Miocene creatures. But curiously, despite the processing of many tonnes of the bone-rich

conglomerate, only a single tooth of *Litokoala kutjamarpensis* has ever been found. Even the major expedition of 1971, led by Drs Bill Clemens and Mike Woodburne, and our subsequent expeditions throughout the 1970s and early 1980s similarly failed to turn up more. As a result, along with a younger species, *Koobor jimbarratti*, this koala is infamous by being known from just a single upper molar.

The koala and its world With just a single molar to go on, reconstruction is nearly impossible. However, its apparent similarity to the modern genus of koalas, *Phascolarctos*, suggests at least something about its structure and from the size of the molar it is possible to say that this species was about two-thirds the size of the living koala. More than that would be rampant speculation.

In life, the Kutjamarpu Local Fauna was, compared to the Ditjimanka and Lake Palankarinna South Local Faunas, quite diverse. It contained about four kinds of dasyurids; about four kinds of bandicoots; a thylacinid

(not yet described); a wombat (a primitive form known as *Rhizophascolonus*); a marsupial lion (a species of *Wakaleo*); an ektopodontid possum; three kinds of ringtail possums; other kinds of possums (yet to be described); a diprotodontid (a species of *Neohelos*) and several kinds of kangaroos.

In addition, there are extinct species of emus, pelicans, ducks, gulls, turtles, snakes, lizards, crocodiles, lungfish and teleost fish. Together these suggest a fauna that was more diverse than the present one, and one that had more modern types of animals in it than the older Ditjimanka Local Fauna from the same general region.

There is nothing here to necessarily suggest rainforest, but the three kinds of ringtail possums plus the koala, all of which are obligate leaf-eating animals, argue for the conclusion that the vegetation at least near the floodplain was dense and diverse in plant species. Perhaps there was gallery rainforest along the edges of what was normally a quiet floodplain in an otherwise open forest habitat.

In terms of the diet of the

in this chapter.

To help visualise the discoveries of fossil koalas that have occurred since 1953 (when the first fossil koala was found), we have also included a summary of the events leading up to the discovery of many of the more interesting specimens.

It is clear from this review that there have been many quite different kinds of koalas and that many of these occupied habitats, such as rainforests, that today's koalas do not seem to be able to colonise.

It is also evident that it has been common in the past for more than one kind of koala to exist at any one time. The modern situation where only one kind of koala survives in what is a relatively limited variety of environments should give us pause for concern. Is it possible that, like thylacinids, the koala lineage is nearing the end of its run?

Koala Relationships

To the palaeontologist and evolutionist one of the most interesting aspects of the fossil record is trying to decide the relationships between the ancient koala

Kutjamarpu koala, it is provocative to consider that the original paper describing the Kutjamarpu Local Fauna reported fossil leaves from the Wipijiri Formation identified as *Eucalyptus* (the word "Kutjamarpu", incidentally, coming from the local Aboriginal dialect, means "many leaves"). This report has since been challenged because it was not based on an analysis of cuticle, something which is important in distinguishing the leaves of this genus from those of other myrtaceous plants.

As a result, it is far from clear when koalas began to focus on *Eucalyptus* as a principle food, although it is probable that *Eucalyptus* had its origins as a rainforest plant and hence probably occurred in what was the ancestral habitat for koalas as a whole. The two groups may, therefore, have coevolved for at least 15 million years.

From life to fossil As noted above, the Wipajiri Formation represents a very distinctive depositional environment. It consists of a heterogeneous pebble conglomerate (a sediment made up of large pebbles as well as fine particles) of varying thickness between about 20 and 50 cm topped by a very fine-grained shale of unknown thickness.

The basal conglomerate appears to represent the base load of a rapidly moving stream. Some of the lumps in it are over ten centimetres in diameter which suggests a reasonable amount of energy in the stream. This in turn suggests either that it was a local flood rush or that the water was moving over a relatively steep surface. Since the whole region is more or less flat-lying, the local flood scenario is more probably correct.

In support of the local flood scenario, we have often noticed that the largest bits of bone and the largest pebbles seem to occur at the base of the conglomerate, suggesting a single flood rush. Further, many large turtle shell fragments are standing on their edge in the sediment, something that would be improbable in anything other than a flood situation.

The conglomerate gives way vertically to a fine-grained leaf shale which contains leaf fossils as well as occasional turtles and fish (all of these fossils being flat-lying). This part of the Wipijiri Formation appears to represent a quiet lake or billabong situation within which only fine silt was precipitating.

The holotype and only known specimen of *Litokoala kutjamarpensis* came from the conglomerate. It is possible that the koala that owned this tooth was sitting in a tree that was knocked over by the flood waters. Or, equally plausibly, the flood waters scoured the surrounding landscape and brought together into the stream's bed load whatever bones and stones lay on the surrounding flood plain. Wear on the tooth indicates that this animal was not elderly, so it presumably died prematurely.

The Kanunka *Litokoala*
Springer, 1987

Age: Approx. 14 million years
Local fauna: Kanunka North Local Fauna
Geological formation: Etadunna Formation
Type locality: UCR Locality RV-8453, Lake Kanunka, South Australia

species. New species found as fossils are certainly fascinating, but their real value is in what they have to tell us about the course of evolution.

In our attempts to unravel this aspect of natural history, we seek to use every piece of information that might have evolutionary implications, although not all information has the same value. For example, experience teaches us that the age of a species is of less importance than its form as an indicator of so-called "primitiveness".

Certainly it is encouraging to know that a species that appears to be the most primitive is also the oldest, but if this is not the case, so be it. It only means that we presumably have a lot more to discover about the fossil record before all of the pieces fit into a single coherent picture.

Speculation about evolutionary history within the koala family was rather pointless before 1957 because until that time there was in fact only a single koala known: the living koala *Phascolarctos cinereus*. Even with the discovery of the Miocene koala, *Perikoala palankarinnica* (about 15 million years old), all that could be said was that there were two

This species was discovered in 1985 by Mike Woodburne, Mark Springer, Judd Case and their colleagues. It was found in a channel deposit developed within the Etadunna Formation.

The koala and its world This koala was about the same size as *Litokoala kutjamarpensis* and hence considerably smaller than the living species. It was also probably very similar to the Kutjamarpu species, differing so far as is known only in small details of tooth morphology.

Nothing is noted about associated species in the original description of the Kanunka *Litokoala*, but judging by its occurrence in the Etadunna Formation, it probably would have been associated with the same sort of faunal elements as the species of *Perikoala*, Mada koalas and possibly *Litokoala kutjamarpensis*. Similarly, until the nature of the associated species is detailed, it is not yet possible to infer the habitat in which this species lived.

From life to fossil As a fossil found in a channel deposit, we may infer that, like *Litokoala*

kutjamarpensis, the Kanunka *Litokoala* fell or was washed into a creek or flood. Because it is represented by only isolated teeth, it may be further inferred that either the skeleton had broken up before being deposited or the first place of rest was subsequently gouged out by another stream and redeposited as small pieces further downstream at the place where it was found by Springer and his colleagues.

The Riversleigh *Litokoala*

Age: Approx. 14 million years
Local fauna: Dwornamor, Henk's Hollow, RSO, Outasite and Gotham Local Faunas
Geological formation: Unnamed freshwater carbonates
Type locality: Riversleigh Station, northwestern Queensland

This species was discovered in 1983, although we did not know that it had been discovered until long after the limestones in which it lay had been dissolved by acid.

The first known specimen turned up in the Gag Site material which was the first of the extraordinarily rich middle Miocene sites (10-15 million

years old) first discovered on Riversleigh Station in 1983. Discovery of Gag Site and of all of the incredible Riversleigh deposits found since that time has almost trebled our entire knowledge of Australia's Tertiary animals (2-65 million years old). At last count, there were almost 200 new kinds of animals recovered from over 50 new local faunas, which variously appear to range in age from about 15 million to about 50,000 years old. These discoveries are, in a very real sense, eclipsing much of what we understood about the history of mammals on this Continent, and they promise to do the same thing for the other groups of Cainozoic vertebrates (less than 65 million years old).

Work at these sites is really just beginning. Each year since 1983 has seen the discovery of many more new kinds of extinct creatures, new local faunas and whole new fossiliferous subregions within the Riversleigh area.

The koala and its world The Riversleigh koala material has not yet been scientifically described, but new and better specimens are constantly accumulating. As this

phascolarctids, one older than the other.

It was not until publication in 1967 of *Litokoala kutjamarpensis* (about 14 million years old) and in 1968 of the more recent *Phascolarctos stirtoni* (between 2 million and 35,000 years old) that potentially conflicting interpretations of their relationships within the family could have occurred.

There was little room for conflict about the relationships of the Pleistocene species, *Phascolarctos stirtoni*, unless it was to doubt whether or not it was sufficiently distinct from the living species to warrant species recognition. It was clearly much more closely related to the living species than to the Miocene *Phascolarctos palankarinnica* or *Litokoala kutjamarpensis*.

The relationships of *Litokoala kutjamarpensis*, however, were not so clear. At the time, it was suggested that this koala was more closely related to the other Miocene than the living species, a conclusion which is now no longer in vogue. In 1978, Archer suggested that it was in fact the most primitive of known koalas, a view based on study of the published diagram of the only known account is being written, a whole upper jaw is emerging from limestone blocks from the Riversleigh Outasite Locality. This will provide the first real understanding of the morphology of the upper toothrow of a species of *Litokoala*, and hence a test of current understanding about the evolutionary relationships of this genus to the other kinds of koalas.

Until the species is scientifically described during the two years to come, we can only say here that it is indeed a remarkable sort of koala. Although it appears likely that it represents an evolutionary link between some of the central Australian Miocene koalas and the genus of living koalas, the emerging upper tooth row suggests that the Riversleigh species of *Litokoala*, at least, had some unique specialisations of its own. Considering that it is probably only a matter of time before the Riversleigh deposits produce a whole skull of this animal, any pronouncements made at this stage are bound to be quickly outdated.

It is easier at this point to provide a picture of the Riversleigh koala's world than it is of the koala itself. From what we presently understand of the sediments and the animal remains recovered from them, when the koala was alive the area was probably covered in rainforest. This conclusion is supported most strongly by the very high diversity of selenodont possums in the Riversleigh middle Miocene local faunas, the number probably being about five.

This very high diversity of obligate tree leaf-eaters means that plant species diversity also had to be very high in order to allow these possums to coexist. High plant diversity enables different kinds of possums to specialise on a narrow range of particular plant types. To be able to do this the plant species utilised must have been predictably abundant otherwise possums specialising on those plants could not persist.

Putting all of this together, we are led to conclude that we are dealing with a dense species-rich rainforest for the middle Miocene world of Riversleigh. In forests of this type, plant species are commonly spread evenly throughout, rather than clumped as in sclerophyll forests, thereby enabling a wider range of herbivorous mammals to coexist in the same region without suffering competition from other species for resources.

From the nature of the limestone sediments and some of the aquatic animals found in them, we have tentatively concluded that during the middle Miocene period at Riversleigh, the region was covered by dense rainforest within which were a myriad of lakes of varying depths filling depressions developed in the older Cambrian limestones (about 550 million years old) that often rose like ramparts among the forest trees.

It is possible that many of the pools were spring-fed, and that these springs built up large hills with pools on their flanks, until the accumulated mass of sediments was so great that the springs had to break out somewhere else, thereby starting another but younger sequence of fossil deposits.

The kinds of animals that provided company for the Riversleigh koala were many. In fact, it would be easier to note

tooth. Having now studied the actual tooth, he no longer holds this view and, in accord with Woodburne, Tedford and Pledge, regards the species of *Litokoala* to be most closely related to the species of *Phascolarctos*.

The description of the two Pliocene species of *Koobor* (between 4 and 5 million years old) marked the beginning of an appreciation of the breadth of the koala family tree. The species of this genus were so unusual that the dissimilarities between the species of *Perikoala* and *Phascolarctos* seemed relatively minor.

But it will not be until late 1987, with publication of three papers describing a total of six new Tertiary koalas, that the complexity of the koala's family tree will really began to be appreciated. Now, together with the Riversleigh finds, we are about to obtain a much clearer picture of the relationships between the present and extinct species of koala.

Constructing a Family Tree

Since Archer's first attempt in 1978 to phylogenetically analyse the interrelationships of the six species of fossil

which kinds of animals did *not* seem to have a representative in these Miocene forests, for this list is very short indeed.

Not so far known to be present at Riversleigh were numbats (the only living species of which is an endangered insect-eating marsupial now restricted to Western Australia, of which there are no known Tertiary fossils from anywhere in Australia), tarsipedids (honey possums, which so far also lack any known Tertiary fossils), echidnas (which curiously have no pre-Pliocene record i.e. no record older than 5 million years old), ektopodontids and miralinid possums (see the the cohabitants of *Perikoala palankarinnica*). Apart from these absences, virtually every other known group of Australian mammals, living and extinct, is well represented in the Riversleigh deposits. In addition, there are many kinds of extinct Riversleigh creatures, such as a very strange new, possibly egg-eating marsupial called "Thingodonta", so far unknown from anywhere else in the world. Riversleigh faunal diversity was extremely high.

Because of the basic

conclusion that the Riversleigh Miocene creatures represent rainforest animals, it is particularly interesting to note that this is the only koala known to have inhabited rainforests. Further, its apparently close relationship to the living genus of koalas suggests that the modern koala had its ultimate origins among the dense rainforests of northern Australia, despite its present total exclusion from rainforests of any kind.

From life to fossil From what we now understand about the lime-enriched waters of the Riversleigh region, the edges of the quiet Miocene lakes were probably covered by mats of vegetation or fine limestone crusts, like ice, which served to trap unwary creatures that ventured out onto their surface.

When a creature broke through the crusted surface, the crust itself settled down to the bottom with the would-be fossil. These crusts often stand out in relief during the acid-etching process that takes place in the laboratory to free the chemically inert bones from the soluble limestone.

Once on the bottom of the Miocene pool, the supersaturated waters began the slow process of fossilisation almost immediately. Today, it is even possible to find cow bones in the Gregory River that are partially encrusted by calcium carbonate. The beginnings of entombment in rock, therefore, evidently require less than a century.

After they were encrusted and subsequently buried by more accumulations of limestone, the molecules of the Miocene bones began an incredibly slow but steady process of exchange with the molecules in the surrounding ground water. Eventually, after many thousands or millions of years, the fossils would have come to represent a balance between original organic molecules and substituted inorganic molecules. The result is a fossil that is now "petrified" and inert to most of the things that would chemically damage a real bone, but one that is so perfectly preserved that examination of its structure under a scanning electron microscope could not reveal how it differed from real bone.

Knowing this, in the not too

koalas then known, essentially only one other novel attempt has been made to reconstruct the koala family tree: the community effort of Woodburne, Tedford, Archer and Pledge in 1987.

The current hypothesis is that the species of *Litokoala* are most closely related to those of *Phascolarctos*; the species of *Perikoala* are most closely related to the combined group *Litokoala/Phascolarctos*; the Mada koalas are most closely related to a combined *Perikoala/Litokoala/Phascolarctos* group; and the species of *Koobor* are the most primitive of all koalas and stand apart from all others which more closely relate to each other.

This sort of relationship statement is called a "cladogram". It expresses hypotheses about evolutionary events that derive from studies of the degree of relatedness between organisms.

This particular cladogram, for example, implies that the species of *Litokoala* and those of *Phascolarctos* shared a common ancestor that gave rise to no other group of koalas. It also implies that the species of *Koobor* represent a specialised group of otherwise primitive koalas that retain, more

distant future we intend to examine the ultrastructure of the tooth enamel of the Riversleigh koala as a possible test of ideas about its evolutionary relationships to other koalas.

De Vis' Mada Koala
Woodburne et al., 1987

Age: Approx. 15 million years
Local fauna: Pinpa, Ericmas and Tarkarooloo Local Faunas
Geological formation: Namba Formation
Type locality: Site D on Lake Pinpa, Frome Downs Stn, South Australia

The first significant discovery of fossil bones in the Frome Embayment region of South Australia was made by geologist Roger Callen. While examining the superficial sediments exposed on the sides of a number of salt lakes on Frome Downs Station, he found pieces of turtle shell and other bones representing previously unknown Miocene animals.

Callen sent these to Richard Tedford who organised an expedition from the American Museum of Natural History in 1971. Members of the expedition included Drs Patricia and Tom Rich (then of the AMNH) and Dr Rod Wells (of Flinders University).

The discoveries of this 1971 expedition were plentiful, including an isolated tooth of what turned out to be a 15 million year old platypus-like monotreme. But in terms of our present interest, their discoveries at Lakes Pinpa and Namba were most significant. Here they found fossil-rich layers of clay containing the remains of many kinds of animals ranging from mouse to cow in size.

During a return expedition to Lakes Pinpa and Namba in 1973, with five Australian colleagues, Tedford discovered a number of excellent koala specimens including the holotypes of De Vis' Mada Koala and Wells' Mada Koala.

The koala and its world De Vis' Mada Koala was approximately the same size as the living koala but was quite different in details of dental morphology. For example, the surface of the last molar was considerably smaller in the fossil

species. What these differences reflect about this Miocene koala's lifestyle is, however, unclear.

The world of this koala was probably similar to that of *Perikoala palankarinnica* and the Robust *Perikoala* which lived at about the same time on the other side of the Flinders Ranges in the Lake Eyre Basin. Many of the same sort of animals lived in the two regions including similar species of browsing ilariids (giant diprotodontan marsupials related to wombats and koalas) and leaf-eating ringtail possums, miralinids, ektopodontids and pilkipildrid possums.

Also similar was the terrain: a large lake, or series of lakes, set in an otherwise relatively flat countryside. The vegetation around the lakes was probably mixed with some rainforest as well as more open forest. None of the animals present was clearly a grazer and hence it is probable that the ground cover was ferns, mosses and sedges rather than grass.

One striking difference, however, was one of the occupants of the lake, a freshwater dolphin. These animals probably first entered the region from the

Some 14 million years ago the now semi-arid Riversleigh region appears to have been covered by closed rainforest, and was the home of the Riversleigh koala, depicted here with a juvenile. A carnivorous marsupial lion and a strange kind of kangaroo are also shown.

The differences between the southern and northern koala races are quite noticeable. The southern race (above) larger with a heavier, shaggy coat and much more fur around the face and ears. The northern koala (right) is smaller and sleeker.

Working in a sea of sand dotted with saltbush, Steven Van Dyke of the Queensland Museum digs for fossil koala bones at Tedford Locality East, Lake Palankarinna in South Australia.

than any other koalas, the greatest number of primitive features within the family.

Athough it is also possible that a species of *Koobor* was ancestral to the rest of the koala tree, this is not a hypothesis inherent in the cladogram, anymore than is the notion that a species of Mada koala was the ultimate ancestor of species of *Perikoala*, *Litokoala* and *Phascolarctos*. What is inherent in the cladogram, however, is the hypothesis that the common ancestor of the koala family tree, based on known species, shared more features in

common with the species of *Koobor* than it did with any other particular known genus of koalas. Similarly, the cladogram suggests the hypothesis that the common ancestor of the species of *Perikoala*, *Litokoala* and *Phascolarctos* more closely resembled the species of Mada koalas than it did the species of the other two genera.

One curious point that arises from this family tree is the prediction that a species of *Koobor* or a *Koobor*-like koala will be found in the Miocene sediments of some area of Australia. At present they are only known from early Plio-

Spencer Gulf to the south but at some pre-Middle Miocene time (before 17 million years ago) in their history became isolated in the Lake Frome region. That they were freshwater dolphins is indicated by the lack of any other marine organisms in the deposits.

From life to fossil The situations in which De Vis' Mada Koala has been found fossilised suggest diverse paths to immortality. In the Ericmas Quarry a lower jaw of this species was found in a stream channel deposit and was probably fossilised in the same way as the tooth of *Litokoala kutjamarpensis*.

In the Lake Pinpa deposits, however, the remains of De Vis' Mada Koala occur in small discrete accumulations much like the teeth of the Robust *Perikoala*. However, these accumulations frequently have a centrally situated white lump that markedly resembles those found in the excreta of living birds and reptiles such as pythons or lizards. Further, the bone-rich accumulations are clearly lense-shaped, and do not appear to be the result of entrapment in a

depression within what was the lake's floor. These observations lend support to the notion that some or all of the accumulations represent the faecal masses of a large Miocene predator or scavenger.

Wells' Mada Koala
Woodburne et al., 1987

Age: Approx. 15 million years
Local fauna: Ericmas Local Fauna
Geological formation: Namba Formation
Type locality: Ericmas Quarry, Lake Namba, Frome Downs Stn, South Australia

The holotype, a right lower jaw with most of its teeth in place, was collected by Rod Wells in 1973, in Ericmas Quarry. This site was not particularly prolific as a fossil producer so the discovery met with considerable excitement. It came at a time when the relationships between the Ericmas and Pinpa Local Faunas were in doubt and promised to help resolve whether there were species-level differences between the two assemblages.

The Koala and its world Wells' Mada Koala was about the same size as De Vis' Mada Koala and hence about the same size as the modern koala. Its teeth, however, were more powerfully built than both De Vis' Mada Koala and the living species, which suggests that it was feeding on different types of leaves. But again, with so little of this animal preserved, it would only be rampant speculation to suggest in what other ways it may have differed from other koalas.

Ericmas Quarry itself almost certainly represents a stream deposit accumulated in a region of sands and clays. Not surprisingly, therefore, buried with the koala were the remains of lungfish, teleost fish, turtles, crocodiles, dolphins and the platypus-like monotreme (*Obdurodon insignis*).

As terrestrial drop-ins with the koala were dasyurids (carnivorous marsupials), ringtail possums and diprotodontids. There was also a specimen referable to De Vis' Mada Koala in the same quarry as the holotype of Wells' Mada Koala. This is of particular interest because it is rare among mammals for two

All that could be said about the evolution of the koala family in 1978 is summarised in (B), depicting possible relationships between the living koala (Phascolarctos cinereus) and the five fossil koalas known at the time. The description of six new fossil koalas has changed the hypothesis markedly. The current hypothesis shown in (A) is that the species of Litokoala are most closely related to the genus of living koalas Phascolarctos; Perikoala species are most closely related to the combined Litokoala/Phascolarctos group; the species of Mada koala lie outside a Perikoala/Litokoala/ Phascolarctos group; and the species of Koobor are the most "primitive" of all koalas, standing apart from the rest.

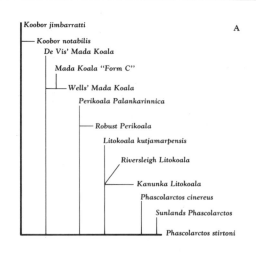

A

Koobor jimbarratti
Koobor notabilis
De Vis' Mada Koala
Mada Koala "Form C"
Wells' Mada Koala
Perikoala Palankarinnica
Robust Perikoala
Litokoala kutjamarpensis
Riversleigh Litokoala
Kanunka Litokoala
Phascolarctos cinereus
Sunlands Phascolarctos
Phascolarctos stirtoni

similar-sized species of the same genus to occur in the same ecosystem. When this sort of thing happens, it is common for competition to occur, resulting either in exclusion or size change in one of the species.

In this case, however, we cannot take association in the stream deposit as evidence of ecological association in life. The stream may have sampled bones or carcasses from a wide variety of habitats before dropping the bones as a single accumulation further downstream.

Taken together, the Ericmas fossils and sediments suggest an environment similar to that which surrounded De Vis' Mada Koala, with the addition of some large streams that probably emptied into the regional freshwater lakes. These substantial stream deposits suggest topographic relief or irregular high rainfall with heavy runoff.

From life to fossil The Ericmas Quarry sediments suggest that some animals associated with Wells' Mada Koala fell into periodically deep and/or fast-moving streams where their bones accumulated as part of the stream's sediments. Ericmas Quarry suggests that some of the material accumulated near a bend in the stream.

Mada Koala "Form C"
Woodburne et al., 1987

Age: Approx. 15 million years
Local fauna: Ditjimanka Local Fauna
Geological formation: Etadunna Formation
Type locality: Tedford Locality, Lake Palankarinna, Etadunna Stn, South Australia

Tedford Locality was first discovered by Dick Tedford in 1953 when he made his momentous discovery of *Perikoala palankarinnica* along the western side of Lake Palankarinna.

From then until the 1970s, it was a locality largely lost sight of in the excitement of new discoveries from other areas. And it was not until it was accidentally rediscovered by Archer in 1971 that it became once again a centre of interest. After preliminary test digs, Woodburne and Archer decided in 1971 to devote 1972 to a major excavation at this site.

As planned, they spent the southern summer excavating and screen-washing bone-rich clays from Tedford Locality. The weather was so hot that the insoluble concentrate, with its masses of fish bones and few precious mammals, dried almost as soon as we spread it out. This meant that we could spend the evenings sorting the concentrate. In this way, the isolated teeth representing Mada Koala "Form C" were found.

The koala and its world This koala was, like the other species of Maka koala, about the size of the living species but had narrower teeth. It also differed from all other koalas in details of tooth morphology. Until, if ever, more material of this species is found, little else about its general appearance can be deduced.

The species that shared its world were the same as those that coexisted with *Perikoala palankarinnica*. In addition, of course, it shared the spotlight with that koala. This is, therefore, a second example of two species of koala occurring in the same fossil deposit. However, in this case there is little reason to doubt

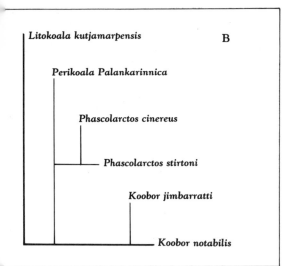

cene sediments but, if the suggested evolutionary relationships are correct, an ancestral form for this otherwise most primitive group of koalas should be found in the fossil record predating the first appearances for the other genera.

This, as fragile is it may be, is the most recent and seemingly most parsimonious interpretation of intrafamilial relationships within the koala family.

Koalas in a Changing World

All koalas so far known are browsing marsupials that probably ate tree leaves.

that the two Tedford Locality koalas coexisted. They differed considerably in size and tooth shape and hence probably were not in competition for the same foods.

It is also probable that the two Tedford Locality koalas were not transported far before burial. Tedford Locality sediments are fine-grained and do not suggest the high stream energy as those from Ericmas Quarry.

From life to fossil The history for the teeth of Mada Koala "Form C" would have been similar to that for the fossil material *Perikoala palankarinnica*.

Koobor notabilis
De Vis, 1889

Age: Approx. 4 million years
Local fauna: Chinchilla Local Fauna
Geological formation: Chinchilla Sand
Type locality: Chinchilla, southeastern Queensland

The collection details for the holotype are unknown. Its describer, the Rev. C.W. De Vis, former Director of the

Queensland Museum, described the collection locality as "a gathering place enriched by agencies of unusual range and efficacy..."

It was suggested by Alan Bartholomai that, on the basis of its preservation, the holotype was collected from Chinchilla, and in support of Bartholomai's conclusion, Archer collected the second known specimen of this species from Chinchilla in 1973; an isolated third premolar.

The koala and its world The species of *Koobor* were radically unlike those of any other koala genus. In fact, there is some doubt that they were koalas at all. It is possible that they represent yet another family of marsupials that merely resembles koalas in their selenodont tooth morphology. Better material will have to be discovered before this question can be resolved. What can be said with certainty, however, is that *Koobor notabilis* was a bit smaller than the modern koala and the species of Mada koala.

The Chinchilla Local Fauna, of early to middle Pliocene age, was very diverse compared to

the central Australian Miocene faunas. It contained large marsupial lions, many wombats, several diprotodontids (two of which were the size of cows), many kinds of kangaroos, including the first known grazing forms, a few possums (probably there were more but they have not yet been recovered) and even a rodent.

Among the important differences in *Koobor notabilis's* world from the previous ones of the middle Miocene were: the abundance of grasses as judged by the presence of grazing kangaroos; the absence of the more archaic groups such as wynyardiids, ilariids, miralinids, pilkipildrids and ektopodontids (see *Perikoala palankarinnica* account); and the presence of rodents (the Chinchilla rodent being among the oldest known for the Continent).

Overall, it is probable that *Koobor notabilis* looked out over a region of open sclerophyll forest fringing an ancestral Condamine River. Almost certainly the dominant trees would have been *Eucalyptus*. The ground would have been covered in grass and shrubs. Large animals would

However, the nature of these leaves and the environments in which they occurred have changed through time just as have the koalas.

It would seem that the oldest koalas known (from the Etadunna and Namba Formations of South Australia) occurred in forested habitats, although there is doubt about the nature of these forests. There is at least some possibility, based on pollens of *Nothofagus*, that rainforest was in the near vicinity. Certainly the evidence from the Miocene deposits of Riversleigh is that that particular species of *Litokoala* lived in lush rainforest.

It is not until we come to the Pliocene koalas that we find them occupying undoubted open sclerophyll forest. Examples of this are the species of *Koobor* from Bluff Downs and Chinchilla. And it is perhaps significant that koalas seem to be absent from the Pliocene rainforest assemblage in western Victoria known as the Hamilton Local Fauna.

In fact, as if to emphasize the point, there is a ringtail possum, *Pseudokoala erlita*, in the Hamilton Local Fauna that seems to have converged on koalas in

have dominated the scene, such as the cow-sized diprotodontids and the many varieties of kangaroos. Leopard-sized marsupial lions would have been hazardous to life, as would an abundance of very large monitor lizards.

From life to fossil The Chinchilla fossils are preserved in a sequence of clays, sands and conglomerates. The greatest variety of Chinchilla species, including *Koobor notabilis* to judge from the isolated tooth found in 1973, are represented in the conglomerate lenses near the top of the middle of the sequence. This particular level appears to represent a series of high-energy flood-plain rather than channel deposits.

It is probable that the Chinchilla creatures either drowned in floods or were accumulated by flood waters. Because most of the Chinchilla fossils consist of isolated fragments of bones or teeth, it is further probable that the bones were tumbled with rocks for some distance before being finally deposited. On the other hand, because whole but fragile skulls are also sometimes found, even of the largest of the Chinchilla mammals (the cow-sized diprotodontid *Euryzygoma dunense*), it is also clear that some of the Chinchilla animals were ultimately preserved very near the site of their death.

Koobor jimbarratti
Archer, 1976

Age: Approx. 4,500,000 years
Local fauna: Bluff Downs Local Fauna
Geological formation: Allingham Formation
Type locality: Scree slope below "Section 1", Bluff Downs Stn, northeastern Queensland

The holotype and only known specimen, an isolated upper molar, was discovered by Archer, Mary Wade, Andrew Elliot and Jim Barratt in 1974. The tooth was in a lump of carbonate cemented sand found at the base of the cliff along Allingham Creek.

Discovery of the Bluff Downs deposit was made by Jim Barratt (formerly of Ayr) while on a fishing trip to Allingham Creek. When he first found bones along the Creek, he dismissed them as probably of little importance because they were not complete skeletons. Sometime later, after he had reported them to Mary Wade at the Queensland Museum, he was surprised at the intense interest his report caused. Until this discovery, there were only four other Pliocene mammal faunas known from the whole of the Australian Continent, and this one promised to be one of the most diverse. What's more, this Bluff Downs fauna was of particular importance because it occurred in sediments that were capped by a basalt that can be dated.

We have made several trips back to Allingham Creek since that time, the most recent being in 1985, and collected more material from what we presume to be the principle bone-producing level in the cliff. Although not all of this material has been processed, so far not another fragment of *Koobor jimbarratti* has been recovered.

The koala and its world We do not know what *Koobor jimbarratti* looked like any more than we know what *Koobor notabilis* looked like. In fact, in

size and some aspects of dental morphology. This would suggest that while koalas per se had abandoned rainforests by this time, there was still a "koala niche" available in these forests that could be filled by an adaptable group of ringtail possums.

In terms of food plants, this transition from rainforest to open forest probably reflects the changes in Australia's climates and vegetation over the same period. As best can be determined, during middle Miocene times (i.e., the time frame for species of Mada koala, *Perikoala* and *Litokoala*), Australia was extensively, though not ubiquitously, cloaked in rainforest.

Then, with the progressive accumulation of water in the polar ice caps and the average drop in world temperatures as the ice ages approached, the Continent began to dry out. It has been suggested that changes in the position of the subtropical high pressure cells over the Continent contributed to a progressive aridity, starting with the southern central region of the Continent.

This drying out probably began in a significant way about 7-8 million years

this case it is even more difficult because *Koobor jimbarratti* is represented by only a single upper molar. Hopefully the material now being processed from the Allingham Formation will produce a more complete specimen.

Fortunately, we know a lot more about the Bluff Downs Local Fauna as a whole. Among the mammals, there were bandicoots, small dasyurids (a species of *Planigale*), at least one kind of ringtail possum, giant wombats the size of wart hogs and large marsupial lions. There were also many kinds of grazing kangaroos as well as browsing species. There were at least four different kinds of large diprotodontids, one of which (a species of *Palorchestes*) may have eaten grasses, bark or some other very tough plant material.

There were also teleost fish, turtles, dragons and large monitor lizards, saltwater crocodiles, another type of crocodile, giant pythons, possible elapid snakes (possibly the oldest record of these poisonous snakes in Australia) and, among birds, the Native Companion (*Xenorhynchus asiaticus*).

Taken as a whole, *Koobor jimbarratti* would have gazed out from his perch in a tree over what was probably a grassed sclerophyll forest or woodland. The most commonly observed animals in this forest were kangaroos. Rarely seen, but constantly feared, were the partially arboreal marsupial lions. At the base of the koala's tree would have been a relatively large body of freshwater containing saltwater crocodiles as well as a very large crocodile of a different kind. As well, there were several kinds of meat-eating turtles. This was not a body of water in which koalas willingly swam.

The Sunlands *Phascolarctos*
Pledge, 1987

Age: Approx. 4-5 million years
Local fauna: Sunlands Local Fauna
Geological formation: Loxton Sands
Type locality: Along the River Murray, 8 km west of Waikerie, South Australia

The local inhabitants of Waikerie, South Australia,

probably knew about the fossils from the cliffs near the Waikerie Pumping Station as early as the 1950s. But the remains of marsupial fossils from these deposits were much slower to turn up.

The only known koala specimen was found here in 1983 by Joe VandeMeer while sieving the Loxton Sands for fossil shark teeth. He had been alerted to the South Australian Museum's interest in marsupial fossils by his friend Jack Maltby, an itinerant fruit picker who spent his summer vacations searching for fossils in the sandy cliffs near the pumping station.

The discovery of the Sunlands Local Fauna was particularly noteworthy because, like the Bluff Downs Local Fauna, it is an important addition to the otherwise poorly-known Pliocene history (5-2 million years ago) of this Continent.

The koala and its world This species has two main claims to fame: it is so far the oldest record for the modern genus of koalas; and, compared to the living species, it is very large.

ago and culminated about 20,000 years ago during the last glacial period.

The shift from rainforest to sclerophyll forest as a koala habitat probably coincided with the gradual expansion of the drier forests and the diminution of the rainforests.

It is interesting to speculate about the possibility that koalas and the genus *Eucalyptus* have been coevolving for at least 14 million years, the age of the oldest claimed association between koalas (*Litokoala kutjamarpensis*) and eucalypt-like leaves. Palaeobotanists have suggested that the genus *Eucalyp-*

tus had its origins in rainforest and only later spread into Australia's expanding arid habitats. In a sense, the eucalypts may have been preadapted to survive in areas of nutrient-poor soil and low rainfall.

If koalas had already become adapted to eucalypts as their particular preferred food tree within the Miocene rainforests, perhaps it was no major break for them to stay with the eucalypts as they were left behind by the retreating rainforests.

But even if this is what happened, why did all koalas vanish from the

Judging by tooth size, it was about 10-20 percent larger than the living species. This size difference qualifies it to be regarded as a giant koala in the same way that large Pleistocene (2 million to 10,000 years old) relatives of living kangaroo species are often referred to as giants. Otherwise, there is little about the teeth of the Sunlands *Phascolarctos* to suggest that it was in any significant way particularly unlike the living species.

The environment of the Sunlands *Phascolarctos* is less easily summarised because it is clear that the site of fossilisation (an oceanic reef deposit) was not the life site for this koala. Its tooth, along with the remains of two other poorly preserved marsupials, was found in a sediment otherwise dominated by shark teeth. Pledge interprets this to mean that a freshwater stream carried the koala tooth out into the marine deposits now known as the Loxton Sands. What type of habitat lined that stream is as yet unknown. The reef site of deposition, however, seems to have been a shallow water oyster bank.

From life to fossil The events leading in this case to fossilisation are equally mysterious. The simplest interpretation might be that a young koala (the tooth is barely worn) lost its grip on a tree branch and fell into the stream below. Here it may simply have drowned and later been carried as a decomposing carcass down to the marine waters of the ancestral Murray embayment. It is also possible that a hur.gry shark ranging upstream in search of food, grabbed the hapless creature as prey when it fell into the creek. As sharks are apt to do, perhaps it dropped a part of the koala which the stream then caught and carried away to the accumulating Loxton Sands.

Phascolarctos stirtoni
Bartholomai, 1968

Age: Approx. 35,000 to 2 million years
Local fauna: Cement Mills Local Fauna
Geological formation: Unnamed cave deposit
Type locality: Cement Mills, near Gore, southeastern Queensland

This koala was found in a lump of rubbly limestone obtained from the Cement Mills limestone quarry. Who found the particular lump and precisely where it came from, if known, has not been recorded.

Encounters with fossil bones in limestone quarries are common. Quarry workers do not want rubbly limestone for their efforts, so when an old cave fill containing bones appears in the quarry wall, the rubbly limestone with its precious load of prehistory is usually discarded.

At least some of the Cement Mills beccia was obtained as discarded blocks near the sides of the limestone quarries.

The koala and its world
Phascolarctos stirtoni was a very modern type of phascolarctid, markedly different from the living species only in its larger size, about 20 percent larger. This larger size is typical of many Pleistocene animals that have living counterparts. The reasons for this are unclear but some are suggested below.

The animals fossilised with *Phascolarctos stirtoni* are a mix of

albeit diminished rainforests? After all, other sorts of possums, such as the group to which the Green Ringtail belongs, managed to hang on in rainforests. We cannot think of a good reason unless competition with ringtails like the Pliocene *Pseudokoala erlita* resulted in their exclusion. If so, why didn't these somewhat koala-like ringtails leave a descendant in the modern rainforests? Again, we don't know. These questions will remain mysteries until the Australian Tertiary fossil record, particularly for rainforest habitats, is a lot better known.

Gigantism and Dwarfism

One final consideration concerns the curious puzzle of gigantism and dwarfism in koala lineages. The oldest-known koalas are all smaller than or equal in size to the living koala. This body size seems to have been a sort of average throughout the known history of koalas, and may reflect some physiologically determined optimal size for a mammal that specialises in eating eucalypt leaves.

The exceptions to this generality are the larger Sunlands *Phascolarctos* from

living and extinct species. Among the extinct are three species of the giant kangaroo genus *Protemnodon*, the marsupial lion, two palorchestids (species of *Palorchestes* which were trunked, cow-sized herbivorous marsupials) and an extinct wombat. Two others now extinct on the mainland, the Thylacine and the Tasmanian Devil, still survived in Tasmania when Europeans arrived. Among the species still living in southeastern Queensland are the Rufous Rat-kangaroo, the Potoroo and the Whiptailed Wallaby.

Taken together, the Cement Mills fossil material suggests to Bartholomai that there was more than one period of time represented by the material. In the main, the faunal material suggests an open sclerophyll forest but several elements suggest

an episode of climatic change that resulted in a brief incursion of rainforest.

Without knowing the precise time within the Pleistocene (2 million to 10,000 years ago) that the Cement Mills local fauna represents, it is impossible to tell whether or not humans ever saw *Phascolarctos stirtoni* as a living animal. The oldest known human remains in Australia are about 30,000 years old, but it is possible that they were in Australia much earlier than this.

From life to fossil During the Pleistocene time, the Cement Mills area undoubtedly had many caves where various kinds of carnivores gathered the remains of animals they had caught or scavenged from the surrounding countryside.

It is entirely possible that the single specimen known of *Phascolarctos stirtoni*, a fragment of an upper jaw, was taken into a cave by one of the larger carnivores such a Thylacine or Devil.

As the material in the cave lair accumulated, carbonate-rich water dripping from the ceiling or flowing over the floor gradually cemented the mixture of pebbles and bones. Eventually, perhaps after thousands of years, the lair would have been filled with this cemented rubble and abandoned by its carnivorous occupants.

From that time on, the rubbly limestone slowly changed and hardened in darkness. Not until quarry explosives tore through the ancient walls of this Pleistocene tomb did sunlight once again fall on the small remnant of this ancient koala.

the Pliocene of South Australia, and the much larger *Phascolarctos stirtoni* from the Pleistocene of Queensland. By late Pleistocene time their probable descendant and certainly only survivor was the smaller, present-day koala, *Phascolarctos cinereus*.

This phenomenon of gigantism and subsequent dwarfing is not unique to koalas. It happened in almost all groups of herbivorous marsupials, and many of these eventually gave way to the smaller but otherwise morphologically similar forms that survive today. The difference in size between the Pleistocene giants and the living dwarfs (or "normals") seems to be proportional to the starting size of the Pleistocene form. For *Macropus titan* (Pleistocene) to *Macropus giganteus* (the modern Eastern Grey Kangaroo), it was close to 30 percent. For smaller wallabies and *Phascolarctos stirtoni* to *Phascolarctos cinereus* it was nearer 10-20 percent.

Why did so many lineages of Australian mammals enlarge up to and during the Pleistocene? And why do so many of these Pleistocene giants seem to be enlarged versions of otherwise similar living species?

One factor that might have contributed to the trend towards gigantism as well as subsequently to the trend towards dwarfism was changes in the Australian climate. With the progressive deterioration of Australia's climates during the last geological era, and the change in vegetation, the nutritional value of an average mouthful of leaves may well have fallen. If so, there would probably have been selective pressure on the herbivores to increase in size because larger herbivores are better able to utilise less nutritious foods (they can eat larger quantities). This selection pressure may well have contributed to the net increase in size in koalas and other herbivores up to

and during the Pleistocene.

But the price paid for this increase in capacity is an increase in the need for a greater amount of water per individual for digestion. By late Pleistocene time, water had probably became a critical limiting factor. Hence the pendulum of selection pressure may well have swung again, with selection favouring animals with a reduced need for free water: dwarfed versions of the Pleistocene giants.

Another possible reason for the drop in size concerns the predictability and length of good seasons. If good seasons stay predictable and maintain a reasonable length, mammals can raise more and/or larger young. But if this predictability falls off and/or the length of the good season shortens, there would be intense selection pressure to reduce the size of the adult in order that the juvenile can be raised through its period of parental dependency as quickly as possible. This is because the duration of parental dependency in animals is normally proportional to the absolute size of the animal. Elephants take much longer to raise their young than do mice.

And suffer the good seasons in Australia did during the more rugged arid phases of the late Pleistocene. Today, long bouts of aridity are one of the principle mortality factors for large kangaroos in dry areas. If the good season ends before the young are weaned, the young die and the maternal investment is wasted.

The historical role of the koala in Australian ecosystems is still poorly known, but with each new koala fossil discovered a clearer picture is emerging. It is hoped that the historical perspective given here may broaden the base on which predictions for the future of the last surviving member of the koala family tree are made.

In many areas the only remaining suitable tree species for koalas are along roadsides. These narrow strips are often essential for their survival, providing the only available corridors between useful habitats.

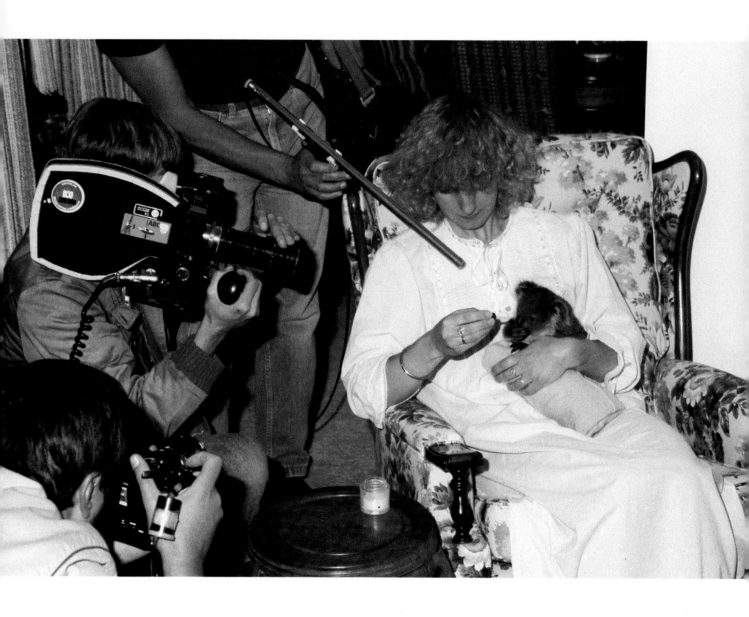

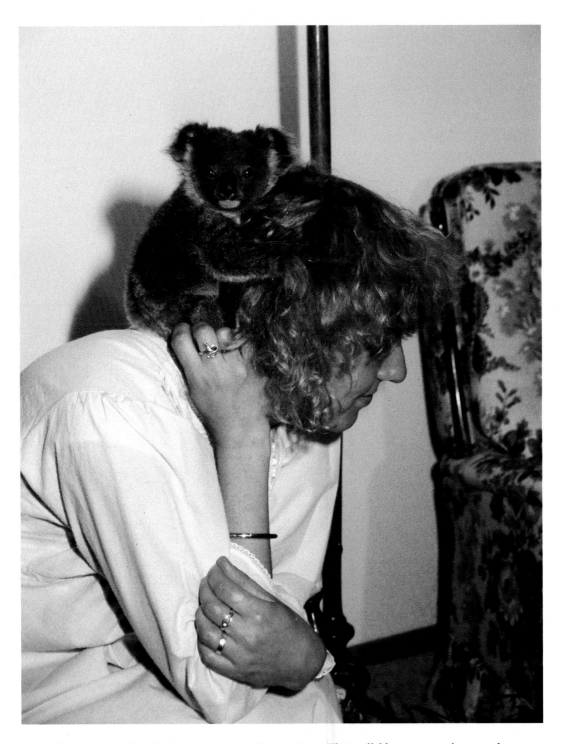

Many orphaned koalas have been raised in captivity. Their affable nature and personality endears them to so many people, here demonstrated by the popularity of this particular infant (left and above).

Of all the Australian animals the koala lends itself particularly well to photography and tourist promotions.

THE KOALA AND MANKIND

Australia has often been called the land of paradox, a term that applies in more ways than one. A little more than 50 years ago many millions of koalas were being killed for their fur, while today, the community at large is both concerned and interested for the future well being of what they see as a very special little Australian.

To trace a relationship between ourselves and the koala, we must go back over 40,000 years to the world of the first Australians. The Australian Aborigines had a special relationship with the land and its animals and plants. According to their beliefs, all life as we know it today can be traced back to the Spirit Ancestors of the Dreamtime, entities in the form of either man or animal, which were possessed with great powers and abilities. The spiritual lives of many modern day Aborigines are still irrevocably entwined with Dreamtime beliefs. While many of these beliefs are known and have been documented by later arrivals to Australia, others no doubt never will be and are already lost forever.

The koala plays an important role in Aboriginal culture, it is connected with several stories relating to the Dreamtime creations, it has a special place in the resting places of spiritual ancestors and there are many legends detailing various aspects of the koala's anatomy and general biology as we see it today. The koala also served as a valuable and much relished food source.

The Dreamtime Creations

The koala first appears in Aboriginal beliefs in stories relating to the arrival of the animals, considered to be ancestral human beings, on the Australian mainland. One charming creation story tells of an island far distant from the continent that is now called Australia, where men who were skilled throwers of the boomerang lived. They loved to engage in contests of skill to show how far or how accurately they could hurl their weapons. As a result of one such contest it was believed that a boomerang had landed on a distant island.

During the ensuing discussions about how best to prove such a claim, a young boy, who had never eaten a koala was given one to eat. He took it to the beach and slit its belly with a flint knife and drew out its intestines. Putting one end of the intestines to his mouth he blew into them until they swelled into a long tube that reached out of sight beyond the curve of the ocean.

The boy's family decided to travel upon this impromptu bridge in the belief that it led to the island beyond the sea. The crossing took many days but eventually they slid down the far end of the arch and found themselves on the island. It was a good place, with green grass, shaded by gum trees and with cool clear water, the island was Australia.

When all the family had come off the bridge they let it float away. The sun shone on it and turned it to many

HOW THE KOALA ARRIVED IN AUSTRALIA

Many Aboriginal groups describe different events of the Dreamtime in relation to their own particular origins, and the following story relates the arrival of the Thurrawal (or Thurrawai) tribe on the coast of New South Wales. This particular version is extracted from *Myths and Legends of Australia* by A.W Reed, published in 1965.

Long before there were men or animals in Australia, the only living things that had eyes to see the vast Continent were flocks of migratory birds. When they returned to their homeland far to the east, they told the animals, which at that time had the form of men and women, of the unending plains, the tree covered mountains, the wide, long rivers, and the abundant vegetation of the great land over which they had flown. The reports created such excitement that the animals assembled from far and near to hold a Corroboree and discuss the matter. It was decided that, as the land appeared so much richer and more desirable than their own, they would all go and live there.

The big problem was how to reach the land of promise. Every animal had its own canoe, but they were frail craft, well suited to the placid waters of lakes and streams, but not to the ocean that lay between the two lands. The only vessel that could contain them all was the one that belonged to Whale. He was asked if he would lend it to them, but he gave a flat refusal.

The animals were determined to migrate, no matter what difficulties had to be overcome. They held a secret meeting at which they enlisted the aid of Starfish, who was Whale's closest friend. Starfish agreed to help, for he was as anxious as the others to make the journey.

"Greetings, my friend," he said to the Whale.

"Greetings," Whale replied in his deep, rumbling voice. "What do you want?"

"There is nothing I want, except to help you. I see your hair is badly infested with lice. I thought that as I am so small I could pick them off for you."

"That's extraordinarily kind of you. They do worry me a bit," Whale admitted. He placed his head in Starfish's lap and gave a sensuous wriggle of contentment. Starfish plucked off the lice in a leisurely fashion.

While the cleaning task went on, the animals went on tiptoe to the shore, loaded all their possessions in Whale's huge canoe, and paddled silently out to sea. The faint splash of their paddles was drowned by Starfish as he scratched vigorously at the vermin.

After a while Whale became restless, and began to fret.

"Where is my canoe?" he asked. "I can't see it."

"It's here, right beside you," said Starfish soothingly.

He picked up a piece of wood and struck a hollow log by his gleaming colours which formed the first rainbow, as it slowly disappeared the boy was turned into a koala and his brother-in-law to a native cat. The other tribesmen remained unchanged and split into a number of groups, each with its own totem, and departed to various parts of the island continent.

Many such stories describe the arrival of Aborigines to Australia, and the origins of some of the animals, including the koala, as we know them today. It must be appreciated, however, that there may have been some details lost during translation and the handing down of the stories from one generation to the next. Similarly, there remain problems of translation and interpretation, not the least of which is the use of the word "koala".

While it is more than likely that koala is an Aboriginal word, it could just as easily be a spelling mistake or one person's interpretation of an entirely different word altogether. The word itself has been variously interpreted as meaning "no drink" or "the animal that does not drink", it has also been interpreted as meaning "biter" or "angry", perhaps a more fitting description after one has had experience in the art of handling wild koalas!

To the Aborigines the koala was an animal known by many different

side. It gave out a booming noise.

"Are you satisfied now?"

Whale sank back again and submitted himself to his friend's attentions once more. The sun was low in the sky when Whale woke up for the second time.

"I am anxious about my canoe," he said. "Let me see it."

He brushed Starfish aside and rolled over so that he could look round him. There was a long furrow in the sand where the canoe had been pulled down to the beach, but of the canoe itself there was no sign. Whale turned round in alarm and saw it on the distant horizon, almost lost to sight. He turned on Starfish and attacked him so fiercely that the poor fellow was nearly torn to pieces.

His limbs and torn flesh were tossed aside contemptuously. His descendants still hide among the rocks and salt water pools as their ancestor did that day, and their bodies bear the marks of the fury of the Whale when he turned against his friend. But little Starfish had not submitted to

punishment without some resistance, and in his struggles he managed to tear a hole in Whale's head, which is also inherited by the descendants of their huge ancestor.

Whale raced across the ocean with water vapour roaring from the hole in his head, and began to overtake the canoe. The terrified animals dug their paddles deeper in the water and strained to make their canoe go faster, but it was mainly through the efforts of Koala that they managed to keep at a safe distance from their infuriated pursuer.

"Look at my strong arms," cried Koala. "Take your paddle strokes from me." The gap grew wider as his powerful arms made the paddle fly through the water, and ever since his arms have been strong and muscular.

The chase continued for several days and nights, until at last land came in sight; the country they had longed for. At the entrance to Lake Illawarra the canoe was grounded and the animals jumped ashore. As they disappeared into the bush the

canoe rose and fell on the waves.

Brolga, the Native Companion, was the only one who had the presence of mind to remember that they would never be safe while Whale was free to roam the seas in his canoe, for at any time he might come ashore and take up the pursuit again. So Brolga pushed the canoe out from the shore and danced and stamped on the thin bark until it was broken and sank beneath the waves. There it turned to stone; and it can still be seen as the island of Canman-gang near the entrance to Lake Illawarra.

Ever since that day Brolga has continued the dance that broke up the canoe.

Whale turned aside in disgust and swam away up the coast, as his descendants still do. As for the animal men, they explored the land and found it as good as the birds had said. They settled there, making their homes in trees and caves, by rivers and lakes, in the bush, and on the endless plains of the interior.

names: Colo, Koolah, Karbor, Cullewine, Boorabee and Goribun to name but a few, primarily as a result of many different Aboriginal dialects present in Australia at the time of European settlement.

It is also of some interest to note that, in view of the often quoted meaning of the word "koala" as the animal that does not drink water, and the established scientific fact that they do, many Aboriginal stories suggest a strong relationship between the two.

It was also suggested by some Aborigines that koalas were the possessors of great magic, including the ability to become invisible at will, as the follow-

ing tale relates:

An Aboriginal man wanted to catch a koala that lived in a large tree. In spite of the objections of his people he took his nullah (club) and climbed the tree, Just as he was about to club the koala, however, the tree opened up and the Aboriginal fell into its hollow interior. No one was game enough to effect a rescue and the unfortunate fellow was left to slowly starve to death.

An interesting conclusion to the above story came many years later when the tree, for its exact location was known to the Aborigines, was blown down; inside the trunk were found the very old bones of an Aboriginal male!

THE WATER-STEALING KOALA

From several Aboriginal tribes come tales of a water-stealing koala, variously named Koobor, Kurburu or Koob-borr. In some versions the koala is in animal form, in others it takes the shape of an orphaned youth. According to legend Kurburu is treated with disdain by the tribe he is associated with. He learned to live off gum leaves but seldom had enough water to quench his thirst. One day when the others were away hunting Kurburu stole all their water containers, hung them in a small tree and climbed into the branches. Using strong magic Kurburu made the tree very tall to escape their wrath.

Upon returning to camp, the Aborigines searched in vain for their water. Sighting Kurburu and their water in the tree they made an attempt to retrieve it but Kurburu was too difficult to shake from his tree, and killed two of his attackers. By calling in clever people, Kurburu was eventually outwitted and hurled to the ground. Once there he received a thorough beating.

It is here that some changes to the story become apparent. In one version the shattered body changed into a koala and climbed into a nearby tree, in others the beating failed to kill Kurburu and he escaped by climbing a very tall tree.

Special laws were passed as a result of this event, and although Aboriginals may kill a koala for food, they must not remove its skin or break its bones until it is cooked. Elsewhere it is stated that the legs must be broken so that the koala cannot escape and take the water with it. It is believed that if these laws are broken then the spirit of the dead koala will cause such a severe drought that everyone except the koalas will die of thirst.

Totems and Culinary Considerations

The koala plays an important role in both traditional and contemporary Aboriginal lifestyles. Roland Robinson, in his 1965 publication *The Man Who Sold His Dreaming*, specifically makes mention of koala totems in talking about the Githavul tribe of northeastern New South Wales. He relates the story of a handsome Aboriginal man, a young girl and the trials and tribulations of their relationship, including interference from "old witches". The story has a happy ending and the family has a spiritual resting place where they are the boorabee or native bears. It is a place where the old people used to go to make lots of boorabee. They used to sing and talk to the stone "Gumbee wanjin boorabee": make lots of bears!

Despite the many tales concerning koalas that can be found in Aboriginal mythology, the apparent laws and customs relating to it, and its totemic role, the koala was certainly a food item much desired by Aborigines and it is often argued that predation by Aborigines was one of the reasons that koala numbers were kept at a relatively low level prior to the arrival of the white man to the shores of Australia. The first documentation of the Aborigines' hunting methods were published in the *Philosophical Transactions* of 1808:

"The New Hollanders (here referring to the Aborigines) eat the flesh of this animal, and therefore readily join in the pursuit of it; they examine with wonderful rapidity and minuteness the branches of the loftiest gum trees; upon discovering the koala, they climb the tree in which it is seen with as much ease and expedition as a European would mount a tolerably high ladder. Having reached the branches . . . they follow the animal to the extremity of a bough, and either kill it with the tomahawk, or take it alive."

Additionally, in 1907 a gentleman by the name of Paul Fountain provided the following account in his *Rambles of an Australian Naturalist*:

"Young koala is much esteemed as an article of food by the blacks, who climb trees and knock mother and young one together from the branches. Though they often fall more than a hundred feet to the ground, they are never killed outright, and sometimes not even disabled. It is left to the knives and hatchets of the gins to complete the cruel tragedy."

The First European Descriptions

Prior to Cook's historic voyage in the *Endeavour*, information gleaned from earlier explorations of the southern oceans had indicated a large southern landmass occupied by animals never before seen by Europeans. To the travellers on Cook's initial voyage to Australia, and to the subsequent parties of convicts and free settlers that followed, the promised land must surely have fostered confusion; it was a largely unknown land, inhabited by "primitive savages" and presenting a bewildering array of wildlife that seemingly defied description. It is perhaps for this reason that many of the early attempts at describing the Continent's unique wildlife, and particularly the koala, show comparisons to animals already known from other parts of the world.

Considering the special role that the koala played in traditional Aboriginal beliefs and culture, it is interesting to note that approximately a decade passed before the koala was brought to the attention of both the scientific community and the white settlers alike. The earliest mention of the koala occurs in 1798 in an article in the *Historical Records of New South Wales*:

"There is another animal which the natives call a Cullewine, which much resembles the Sloths of America . . . "

Some four or five years later a further description was provided by a young naval Ensign, F. Barrallier, as a result of expeditions into the interior of the Continent. Barrallier recorded:

"Gory told me that they had brought in portions of a monkey, but they had cut it in pieces, and the head, which I should have liked to secure, had disappeared. I could only get two feet through an exchange which Gory made for two spears and one tomahawk. I sent these two feet to the Governor preserved in a bottle of Brandy."

Some time later Barrallier obtained a live "monkey", and what is probably the first published account appeared in the *Sydney Gazette* on the 21st of August, 1803:

"An animal whose species was never

before found in the colony is in the possession of His Excellency. When taken it had two pups, one of which died a few days since. This creature is somewhat larger than the waumbut . . . the graveness of the visage . . . would seem to indicate a more than ordinary portion of animal sagacity, and the teeth resemble those of a rabbit. The surviving pup generally clings to the back of the mother, or is caressed with a serenity that appears peculiarly characteristic; it has a false belly, like the opossum, and its food consists solely of gum leaves, in the choice of which it is excessively nice."

And so it began. As information became more readily available so conjecture became more apparent (perhaps the true nature of science) and the descriptions became more fluid in nature. The following extract was was originally published in a newspaper of the early 19th century:

"They are called by some, Monkeys, by others, Bears, but they by no means answer to either species. I first took them to be a species of the Sloth or Buffon, and so they might be, though they differ also in many respects from that animal; and I now think that these animals mostly resemble, and come nearest to, the Loris, or slow paced Lemur of India . . . "

It is now widely known that the koala is neither sloth nor monkey and it is certainly not a "Native Bear" to paraphrase an often used and sometimes generally accepted name. The koala is quite simply a marsupial, but, while some of the blame for the use of the word "bear" must rest with the early naturalists, it is also irrevocably entrenched in the annals of scientific literature. For, like all known plants and animals, the koala is blessed with a scientific name to distinguish it from all others.

In 1814, as a result of a visit to the British Museum, a French naturalist coined the name *Phascolarctos* to describe the genus, and some years later the name *cinereus* was derived, by yet another naturalist, to indicate the species. Thus we have *Phascolarctos cinereus*, a name which, perhaps unfortunately, can only be translated as the "ashy coloured" or "grey coloured" *cinereus* "pouched bear" *Phascolarctos*. The laws of Binomial Nomenclature dictate that this name, however inaccurate, is virtually impossible to change.

The Hunters and Protectors

Given the relatively high visibility of koalas when compared to most other arboreal marsupials, and also the fact that koalas not infrequently come to the ground in order to travel from one tree to another, it is likely that their numbers were kept at a low level by the Aborigines prior to white settlement. Even the famed naturalist John Gould found the koala very difficult to find in the wild and, in what would appear to be a highly prophetic sense, held grave fears for the future of the species over the long term.

With the coming of Europeans to the shores of Australia, and the subsequent displacement of Aborigines, it is logical to assume that the koala popula-

HOW THE KOALA LOST HIS TAIL

This tale is yet another which reinforces a strong association between koalas and water. Here, Koala and Kangaroo are slowly dying of thirst, but, rather than sit and wait for death, they decide to set out across the plains in search of water. Their trek is long and arduous and along the way they pass the remains of many other animals that have already perished in the drought. Finally they arrive at a dried up river bed; Kangaroo suggests that they dig in the river bed for water. Koala is too tired for such strenuous activity and decides to sleep while Kangaroo digs.

After digging an immense hole Kangaroo is exhausted and returns to where Koala is resting, hoping to be relieved. Koala asks if any water has been found yet. When Kangaroo replies in the negative, Koala says that he is not ready and asks to be left alone. Because of Koala's small size Kangaroo takes pity on him and returns to work on the hole. After further digging, his efforts are rewarded and water begins to trickle into the hole. Excited, he races back to Koala to tell him the good news and offer to bring him some water. Koala, who was only pretending to be asleep, springs to his feet, knocks Kangaroo down and races to the water hole where he drinks greedily.

Kangaroo is taken aback by the selfish nature and attitude of his one time friend. Taking out his knife, he creeps up behind Koala who was still drinking from the hole. Koala's long tail was stretched out behind him and with one stroke of his knife, Kangaroo severs the tail at the base. Koala screamed and turned around to see Kangaroo holding the knife in one hand and his tail in the other; he turned and ran into the surrounding bush.

In another legend, again during a drought, the animals noticed that Koala never seemed to suffer from thirst and they all believed that a secret water supply was being kept. Even though a watch was maintained day and night, Koala's secret remained just that until Lyre Bird saw him hanging upside down from the branches, using his strong tail. Curious as to why Koala would hang in such a strange manner, Lyre Bird crept up and saw Koala drinking water that had collected in a fork of the tree.

Believing that the tree might be hollow, and consequently filled with water, Lyre Bird went back to his camp and returned with a firestick, setting the tree on fire with spectacular results. The tree burst and was indeed filled with water, enough in fact to satisfy the thirsts of all the animals. The results of the fire however, did leave their mark. To this day, if you look closely at the tail of a lyre bird, the brown scorches made by the firestick are still evident. Similarly, the koala's tail was burnt off by the flames of the fire; he escaped by scrambling to another tree but for ever after had to live without a tail.

tion of Australia began to increase. Some indication as to the extent of this population increase was given by Parris who, in 1948, described changes in population densities in the Goulburn River district of Victoria over the many decades following settlement. Initially rarely seen upon settlement, koalas were present in their thousands within a short time of the local Aboriginal communities being displaced.

Given such rapid population increases over the natural range of the koala as European settlement proceeded, and the desire of the new settlers to maximise whatever opportunities were presented to them, it was not long before koalas were being shot for sport. By the late 19th century a flourishing fur trade had begun.

Over a 50 year period beginning in the late 1870s, many millions of koalas were killed for their fur, said to be very durable and a particularly effective insulator against the cold. In the midst of this onslaught, even as early as 1898, legislation was passed in at least one Australian State in an attempt to protect the koala from further persecution.

On the 1st November, 1906, the koala received additional protection by virtue of the Native Animals Protection Act. It was not a blanket protection,

Koalas make easy targets for hunters, and during 1924, at the height of the killing, more than 2 million koalas were slaughtered for their fur.

however, and the legislation allowed for open seasons to be declared. It is unlikely that these open seasons were considered in the light of any valid biological reasoning, rather they were there as a loophole for exponents of the fur trade and for farmers needing to rid their newly won lands of the "vermin" they believed infested them. In one year alone, 1924, over two million koala pelts were exported from Australia, and this was when some states were already recognising the koala as an endangered species. Culling was being undertaken on a grand and virtually uncontrolled scale.

There are many heart rending stories relating to the culling, on what has justifiably been viewed by some as the indiscriminate slaughter of the koala in

Australia for the monetary gain of a few. It was also widely accepted that the koala was not an easy animal to kill, not on humanitarian grounds, but because of its tree dwelling habits and thick hide. One hunter, quoted in Charles Barrett's *Koala*, stated: "It is cruelty to shoot at them with shot, if they are any height up a tree; but a bullet brings them down 'by the run'."

Given our current knowledge it would have been a relatively easy matter to predict that the harvesting of koalas at such a rate could not continue indefinitely, and most certainly it did not. By the mid-1920s, grave fears were already being expressed for the koalas' safety, and in some states it was already recognised as an endangered species. The last "open season" on koalas took

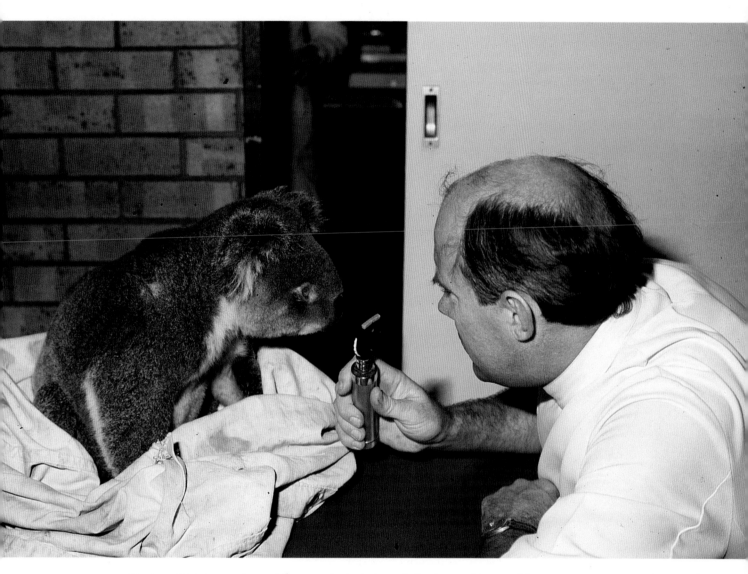

Conjunctivitis is a major problem, particularly in male koalas, and is one of the two major clinical disorders associated with the Chlamydia infections currently affecting large numbers of wild koalas.

place in Queensland over August in 1927. In the short space of just 31 days, 584,738 koala skins were sold, earning for the State Government of the time a princely sum of just under 19,000 pounds in royalties. In what must now appear to be the epitome of hypocrisy, public outrage as a result of the open season was such that, in order to appease the electorate, the Government placed the royalties into a trust fund for the protection and propagation of native fauna!

The koala became fully protected by law throughout all the eastern Australian States in the 1930s.

It has also been argued that the decline of the koala was not solely attributable to the impact of the fur trade. Major disease epidemics, initially in the periods 1887 to 1889 and 1900 to 1903, and again during the late 1920s to early 1930s have also been blamed, along with fires and the clearing of large areas of viable koala habitat for agriculture and urban development.

Shortly after the First World War, koalas were believed to be extinct in South Australia, and with the exception of captive animals, they were also thought to have disappeared from New South Wales at about the same time. In 1934 the Chief Inspector of Fisheries and Game in Victoria, Mr F. Lewis, stated:

"Now the species is almost extinct on the mainland, a very few koalas survive in the Inverloch district and in south Gippsland around Welshpool, Toora, Foster, etc., others are living, and I am glad to say, thriving, on the islands of Western Port Bay. I estimate that there are now not more than 1000 koalas in this State . . . On the mainland of Victoria, I feel certain, the koala is doomed to early extinction, and will never be re-established . . . "

It was sentiments such as these, coupled with an obvious scarcity of koalas within their former range, that led many people to believe that the koala was virtually poised on the edge of extinction throughout Australia. Regardless of whether this was a valid claim or not, there are no grounds for arguing that the preceding 40 or 50 years of koala "management" had not laid the grounds for genuine concern.

From a purely demographic point of view, and with the possible exception of the original South Australian population, it is extremely unlikely that koalas were ever extinct in any of the eastern states of Australia. In accepting the South Australian extinctions it must be recognised as a political extinction only, the former range of the koala certainly had no respect for state boundaries. Rather it was a population at the extreme southern limit of its natural distribution and bound, by virtue of this fact, to be at less than optimum population densities. Even so, its extermination was, and continues to be, a regrettable event which should never have happened; the loss of information from a biogeographic and genetic point of view is significant.

In New South Wales, where extinction was also deemed likely, it must have been plainly apparent that the koala populations were severely depleted. But whether or not they were actually extinct can be subject to debate. Most certainly their numbers would have been reduced to very low levels in areas within reach of humans, and there is a high probability that localised "extinctions" were commonplace. However, there were also substantial areas of virtually untouched lands, primarily focused upon the hinterland and mountains of the Great Dividing Range, that would have contained widely dispersed but viable populations. Indeed, it was more than likely that such populations provided the stock for those now found along the coast of this particular State.

It is one of the strange paradoxes of human nature that, while elsewhere on the Australian mainland koalas were being persecuted for the fur trade, the seeds of conservation were already being sown. To begin with, it must be said that the koala's saviours were no more than private individuals who initiated translocation programmes of koalas to Phillip and French Islands in Victoria in the 1870s. This was a move that proved to have significant consequences for koala management throughout Australia. While the Phillip Island population was slow to increase in the early stages, on French Island by the 1920s there was concern being expressed for the overbrowsing of food tree species by a rapidly growing population. In 1923 a series of transfers were undertaken by Government authorities to other nearby islands and back to the Victorian mainland.

It was recently estimated that over 10,000 koalas have been successfully relocated to nearly 70 localities on the Victorian mainland. These koalas and their progeny, in some cases many generations removed, now occupy many of the areas that they formerly used to occur in quite naturally.

In South Australia similar foresight saved the dwindling populations.

While koalas were generally regarded as being extinct in that State by the early 1930s, in December of 1923 a small number from Victoria were released in the Flinders Chase on Kangaroo Island. Despite a slow start, within a couple of decades they too had increased dramatically and were subsequently used to successfully colonise areas on the South Australian mainland.

In considering the obvious success of the translocation programmes it must be accepted that there have also been failures as well, more often than not the relocation programmes have been motivated by a desire to reduce overcrowding and defoliation of food trees rather than a need to colonise new areas. Fortunately the failures have been few and there is little need to dwell on the morbid aspects in the light of the above.

In several instances the koala has shown itself to be remarkably adaptable. Western Australia, for example, an area of Australia not known to have ever had naturally occurring koala populations, now has a resident koala population in the Yanchep National Park. The colony has taken some 40 years of fostering, but it has been achieved with a very low mortality rate. Similarly, ten koalas from Phillip Island were transferred to a mainland nature reserve without loss; while there may appear to be nothing unusual about this particular translocation, it is of special interest in that their previously known food trees were entirely absent from the area.

The story of the koalas' return from the verge of extinction is often flouted as one of the few Australian conservation success stories. By and large, on a collective scale, there has been no other effort to compare with it, But, by no means wishing to detract from its success, it was due, for the most part, to an extremely fortuitous set of circumstances. The islands off the coast of

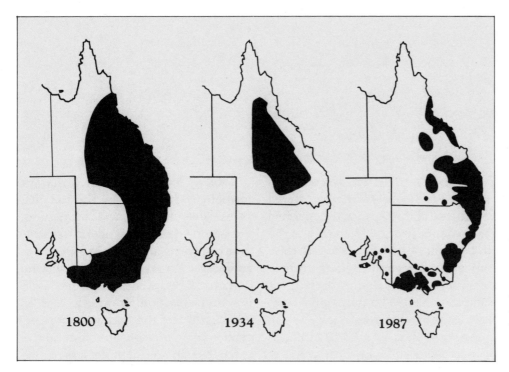

The distribution of koalas in eastern Australia from the time of European settlement to the present day, showing the drastic reduction in numbers until their protection in the 1930s.

Victoria and South Australia did not, at least in recorded European history, already have natural koala populations, and it was little more than luck that there were sufficient trees of the right species available to not only nourish the recently introduced koalas but to also nurture a rapidly growing population.

There were some notions in the early 20th century that the captive propagation of koalas was one way of replenishing the seriously depleted wild populations. While this concept in itself had considerable merit, if purely from an idealistic point of view, it is extremely unlikely that it would have been successful. Given the known reproductive rate of healthy adult koalas, it would only have been through the confinement of large numbers that any degree of success could have been assured. Even then, it would be unlikely that surplus stock could be produced at a rate necessary to override the natural and induced mortality rates that would arise in a newly established "natural" colony.

The Koala in Captivity

The Koala is an animal much coveted by zoological institutions and wildlife parks which, at least in this modern age, display the animal under the auspices of education and conservation. Most certainly in the earlier days, it was more than likely displayed on the basis of being a zoological oddity, yet another from that strange land, Australia.

Early records indicate that the first live koala to reach Europe did so in 1880, surviving what could only have been an arduous sea journey on a diet of dried gum leaves. The animal was purchased by the London Zoological Society and, perhaps surprisingly, considering the then largely unknown nature of the koala's feeding habits, survived for a considerable length of time before succumbing to an unfortunate accident.

The popular image of the koala as a cuddly, loveable animal, along with its "teddy bear" visage has no doubt done much to make it a desirable and popular addition to any wildlife collection both in Australia and overseas. It is an image that has been actively fostered by many exhibitors, and additional fuel was no doubt provided by many of the early accounts which described their captive mannerisms in purely anthropomorphic terms. Charles Barrett, for example, in his 1943 publication *Koala*, provided the following:

"An expression so mild and appealing, and quaint, a Teddy Bear's face, as you see it in some of our pictures, is that of a cherub in fur."

There are many accounts of orphaned koalas that have been raised in captivity and, to provide some defence for the above statement, there would be many zoo keepers, biologists and dedicated individuals who have been fortunate enough to have experienced a young koala's affable nature and singularly peculiar personality that allows it to endear itself to so many.

The koala is, however, a notoriously difficult animal to maintain in captivity. It is a highly specialised marsupial that requires a lot of attention, the obtaining of food for captive animals in itself is a labour intensive and therefore costly undertaking. Many zoos, in fact, solicit sponsorships of individual animals ranging from frogs to elephants as a means of raising revenue and also as a public relations venture. Some indication of the koala's maintenance requirements in this regard can be gleaned by making an enquiry at those zoos currently undertaking such schemes, costs will vary between several hundred and several thousand dollars per individual koala.

Captive husbandry of koalas is further disadvantaged by the fact that they suffer from a wide variety of debilitating disorders while at the same time displaying an apparent aversion to antibiotics. Koalas also stress easily and do not respond well to constant handling and attention. In short, it is an established fact that old age is a rarely cited cause of death amongst captive specimens. In several of the less reputable Australian wildlife parks, perhaps more so in the past than in the present, the death of captive animals has simply been a matter of course, with the maintenance of koalas for display purposes being achieved with the continual replacement of animals from wild populations.

Despite the obvious difficulties of maintaining captive colonies, the koala remains an animal much in demand. With the recent lifting of a long standing Government ban on the export of koalas from Australia, many koalas have made their way overseas to several zoological institutions. The Japanese Government, for example, recently invested millions of dollars in an attempt to establish a captive colony of koalas, yet already there have been some mortalities.

While there are many valid reasons in specific instances for the maintenance of captive koala colonies, there are many who, having witnessed first hand the behaviour and lifestyles of wild koalas, or for otherwise purely ideological reasons, do not wish to see this practice continue.

Towards a Brighter Future?

Primarily as result of those early translocation programmes, but also a reflection of the recovery of natural populations, the koala today can be found over much of its former range. That it will ever occupy the whole of these areas is very doubtful, mainly because of restrictions now imposed by habitat availability. By the willing hand of man the koala has both suffered and been assisted, and by a strange twist of fate, now enjoys complete protection almost to the extent of reverence. The koala has become a national symbol, the focal point of international visits, children's books, advertising and dedicated conservation groups.

In the short space of some 50 years koala populations have obviously recovered, albeit with some help from man, from an onslaught of major proportions. While we can now look back and appreciate what may have happened if events did not proceed the way they did, it is a sobering thought to realise that a very similar thing is happening at this very moment with the kangaroo, the only difference is the regard with which each of the two species is held by our society.

One of the problems that must now be faced in view of the limited areas now available to the koala for colonisation, and the already widely recognised problems in areas where the koala has nowhere else to go, is how best to manage the species for our mutual benefits, for that is the legacy that has been left to us; and that in itself is a very difficult task for any one conservation body or government authority.

What, for example, will happen if overcrowding on the islands off the coast of Victoria, and the many "islands" now created on the mainland as a result of development and the clearing of koala food trees, is left to run its course? Under natural conditions, and removed from the influence of man, the populations would possibly crash through starvation and disease, leaving only the fittest to survive. This is a factor that future managers of the koala will undoubtedly have to consider in specific situations.

Before any attempt can be made to manage koala populations on a national scale, and despite the fact that there is a considerable amount of information available to the scientific community, much more remains to be known about the koala, particularly in the areas of nutrition and population ecology. There are even several aspects of the koala's general biology that remain to be accurately determined.

Many years ago it was suggested that fires played a major role in the decimation of koala populations, particularly in southeastern Australia. Many of us can remember only too well the devastation of the Ash Wednesday fires of 1983, and there are many volunteer and professional bush firefighters who can fully appreciate the impact of wildfire upon koala populations.

Given the gradual evolutionary development of the Australian continent and its unique biota, many of the plants and animals have adapted by various strategies to a life cycle based on fire, indeed it is a necessary part of their life cycle. Several plant species, including some eucalypts, require high intensity fire to release their seeds and thus ensure the propagation of their species. A number of small mammals and some bird species, many of which are in a far more endangered state than the koala, require the frequent burning of their chosen habitat to maintain viable population densities.

As a consequence it must be realised that both the exclusion of, and the too frequent occurrence of fire in a given area can have significant impacts upon the natural communities. The koala is, by all accounts, an animal with a very low tolerance to fire, and in our management of the species' future we need to address this problem in particular, while considering the cost to other plants and animals in terms of maintaining biological diversity.

There are many who would argue that we have come to the crossroads in our relationship with the koala. On one hand we appear to have re-established the koala over much of its former range with some areas in dire need of management because of overpopulation and limited food resources. On the other hand, it has been suggested that the species is yet again to suffer another onslaught, this time from disease.

Much has been said over the last few years of a microscopic organism, a bacterium called *Chlamydia psittaci* (Klam-iddya sit-a-ky). While this organism has been known to be associated with a specific koala disorder for some

time, it is now being suggested that it is in fact singularly responsible for a wide variety of diseases including conjunctivitis, "wet bottom" or "dirty tail" syndrome, pneumonia, and in the females, cystic ovaries resulting in long term infertility. It has also been suggested that this organism has the potential to dangerously deplete existing colonies of koalas perhaps, once more, to the point of extinction.

The reasons behind the increase in the incidence of this disease are still not clear, although one hypothesis widely proclaims that stress is a major factor, brought about by habitat destruction, trauma and overcrowding. This appears to be the case in certain instances, but in others it is not necessarily applicable. For example, the records of the Koala Preservation Society of New South Wales show that, over the course of a 12 month period, there was a significant rise in the number of koalas affected between September and January. This would appear to indicate that the incidence of the disease can also be related to the stress of the reproductive effort.

Also of interest is that, of the two major clinical disorders generally associated with *Chlamydia*, namely conjunctivitis and the dirty tail or wet bottom syndrome, conjunctivitis tends to occur in a significantly higher proportion of males than females. Conversely, the wet bottom syndrome tends to occur in a significantly higher proportion of females.

We need to know much more about the disease problems of koalas before jumping to any rash conclusions. With regard to the problems of chlamydial infection, we need to establish beyond doubt that the diseases are being caused by the one organism, how it is transmitted between individuals, and what makes one koala susceptible yet not another.

We need to know if the koala's immunological defence system is capable of fully protecting it from disease, and if not then why not. Perhaps most importantly, we need to know if the disease and its cycle within koala populations is a natural phenomenon, and should we interfere? Even if we choose to, it has been established that the antibiotics necessary for the control of associated diseases in humans can prove lethal to koalas. By a similar token it would be a herculean task, if not impossible, to safeguard and inoculate wild koala populations against the disease, even if a cure was discovered.

In short, there is much more to learn, and we, as Australians, have been given the task of making the decisions that will determine the future status of the koala in this country. The decisions will need to be made on the basis of our knowledge and understanding, not on conjecture or grandstanding. Our relationship with the koala has survived many thousands of years in one form or another. Over the last two hundred years it could be said that we have passed the lowest point of that association, let us hope that the future brings only improvements.

BIBLIOGRAPHY

INTRODUCTION TO THE MARSUPIALS

Anderson, S. and Knox Jones, J. (1984). "Orders and Families of Recent Mammals of the World". John Wiley and Sons, New York.

Archer, M. and Clayton, G. (1984). "Vertebrate Zoogeography and Evolution in Australasia". Hesperian Press, Carlisle, Western Australia 57, pp 423-486.

Dawson, T.J. (1983). "Monotremes and Marsupials, The Other Mammals". Edward Arnold, London.

Green, B. (1984). Composition of milk and energetics of growth in marsupials. In "Physiological Strategies in Lactation", Eds M. Peaker, R.G Vernon and C.H. Knight. Symposia of the Zoological Society of London No. 51, pp 369-387. Academic Press, London.

Hume, I.D. (1982). "Digestive Physiology and Nutrition of Marsupials." Cambridge University Press, Cambridge.

Lee, A.K. and Cockburn, A. (1985). "Evolutionary Ecology of Marsupials". Cambridge University Press, Cambridge.

Russell, E.M. (1982). Patterns of parental care and parental investment in marsupials. Biological Reviews 57, pp 423-486.

Russell, E.M. (1984). Social behaviour and social organisation of marsupials. Mammal Reviews 14, pp 101-154.

Stonehouse, B. and Gilmore, D. (1977). "The Biology of Marsupials". The Macmillan Press Ltd, London.

Strahan, R. (Ed) (1983). "The Australian Museum Complete Book of Australian Mammals". Angus and Robertson, Sydney.

Tyndale-Biscoe, H. (1973). "Life of Marsupials". Edward Arnold, London.

Tyndale-Biscoe, H. and Renfree, M.B. (1987). "Reproduction in Marsupials". Cambridge University Press, Cambridge.

Whitley, G.P. (1970). "Early History of Australian Zoology". Royal Zoological Society of New South Wales, Sydney

FORM AND FUNCTION IN THE KOALA

Bergin, T.J. (1978). "The Koala". Zoological Parks Board of N.S.W., Sydney.

Cork, S.J. (1984). Utilization of Eucalyptus foliage by arboreal marsupials. Proceedings of the Nutrition Society of Australia 9, pp 88-97.

Cork, S.J. (1986). Foliage of Eucalyptus punctata and the maintenance nitrogen requirements of koalas, Phascolarctos cinereus. Australian Journal of Zoology 34, pp 17-23.

Cork, S.J. and Hume, I.D. (1983). Microbial digestion in the koala (Phascolarctos cinereus) a folivorous, arboreal marsupi-

al. Journal of Comparative Physiology 152, pp 131-135.

Cork, S.J., Hume, I.D. and Dawson, T.J. (1983). Digestion and metabolism of a natural foliar diet (Eucalyptus punctata) by an arboreal marsupial, the koala (Phascolarctos cinereus). Journal of Comparative Physiology 153, pp 181-190.

Cork, S.J. and Warner, A.C.I. (1983). The passage of digesta markers through the gut of a folivorous marsupial, the koala Phascolarctos cinereus. Journal of Comparative Physiology 152, pp 43-51.

Degabriele, R. and Dawson, T.J. (1979). Metabolism and heat balance in an arboreal marsupial, the koala (Phascolarctos cinereus). Journal of Comparative Physiology 134, pp 293-301.

Degabriele, R. (1978). Water metabolism of the koala. In "The Ecology of Arboreal Folivores" ed G.G. Montgomery, pp 163-172, Smithsonian Institution Press, Washington, D.C..

Eberhard, I.H. (1978). Ecology of the koala, Phascolarctos cinereus (Goldfuss) Marsupialia, Phascolarctidae, in Australia. In "The Ecology of Arboreal Folivores" ed G.G. Montgomery, pp 315-327, Smithsonian Institution Press, Washington, D.C..

Hindell, M. (1984). The feeding ecology of the koala, Phascolarctos cinereus, in a mixed Eucalyptus forest. Master of Science thesis, Zoology Department, Monash University, Clayton, Victoria.

Iredale, T. and Whitley, G. (1934). The early history of the koala. Victorian Naturalist 51, pp 62-72.

Lanyon, J.M. and Sanson, G.D. (1986). Koala (Phascolarctos cinereus) dentition and nutrition. I. Morphology and occlusion of cheekteeth. Journal of Zoology, London, 209, pp 155-168.

Mackenzie, W.C. (1919). "The Genito-urinary Tract in Monotremes and Marsupials". Australian Institute of Anatomical Research, Melbourne.

Martin, R.W. and Lee, A.K. (1984). The koala, Phascolarctos cinereus, the largest marsupial folivore. In "Possums and Gliders" eds A. Smith and I. Hume, pp 463-467, Surrey Beatty and Sons, Chipping Norton, N.S.W.

Pratt, A. (1937). "The Call of the Koala". Robertson and Mullins, Melbourne.

Strahan, R. (1986). Black man, white man, koala. Wildlife Australia, Autumn 1986, pp 22-25.

Troughton, E. le G. (1967). "Furred Animals of Australia". Angus and Robertson, Sydney.

Whitley, G.P. (1975). "More Early History of Australian Zoology". Royal Zoological Society of New South Wales, Sydney.

BEHAVIOUR AND ECOLOGY

Bergin, T.J.(ed.) (1978). "The Koala." Proceedings of the Taronga Symposium on koala biology, management and medicine. Zoological Parks Board of N.S.W..

Bollinger, A. (1962). Gravel in the caecum of the koala *Phascolarctos cinereus*. Aust. J. Sci. 24, pp 416-17.

Briese, D. (1970). The oestrous cycle of the koala *Phascolarctos cinereus*. Hons. Research Project, Dept. Zoology, Univ. of Adelaide.

Campbell, P., Prentice, R. & McRae, P. (1979). Report on the 1977 koala survey. Wildl. in Aust. 16 (1) pp 2-6.

Degabriele, R. (1973). Koalas thrive in a tropic haven. Habitat 1 (2), pp 8-11.

Eberhard, I.H. (1972). Ecology of the koala, *Phascolarctos cinereus* (Goldfuss) on Flinders Chase, Kangaroo Island. Ph. D. thesis, Univ. Adelaide.

Eberhard, I.H. (1976). Ecology of the koala, *Phascolarctos cinereus* (Goldfuss) (Marsupialia: Phascolarctidae) in Australia. in "The ecology of Arboreal Folivores", G.G. Montgomery (ed.), Proc. Nat. Zool. Park Symp. No.1, Smithsonian Institute Press, Washington, D.C.

Fleay, D. (1937). Observations on the koala in captivity, successful breeding in Melbourne Zoo. Aust. Zool. 9, pp 68-80

Fleay, D. (1984). Koala twins one too many. Courier-Mail (Brisbane) 11 Dec. 84.

Gall, B.C. (1978). Koala distribution in New South Wales. in "The Koala" ed by T.J. Bergin. Taronga Park Zoo, Sydney, p 115.

Gall, B.C. (1978). Koala research, Tucki Nature Reserve. in "The Koala" ed by T.J. Bergin. Taronga Park Zoo, Sydney, pp 116-24.

Gall, B.C. (1980). Aspects of the ecology of the koala, *Phascolarctos cinereus* (Goldfuss), in Tucki Tucki Nature Reserve, New South Wales. Aust. Wildl. Res. 7, pp 167-76.

Gordon, G. & McGreevy, D.G. (1978). The status of the koala in Queensland. in in "The Koala" ed by T.J. Bergin. Taronga Park Zoo, Sydney, pp 125-31.

Kershaw, J.A. (1934). The koala on Wilson's Promotory. Vict. Nat. 51, pp 76-7.

Kikkawa, J. & Water, Margaret (1968). Report on the koala survey, 1967. Wildl. in Aust. 6, pp 100-103.

Lewis, F. (1954). The rehabilitation of the koala in Victoria. Vict. Nat. 70, pp 197-200.

Marlow, B.J. (1958). A survey of the marsupials of New South Wales. C.S.I.R.O. Wildl. Res. 3, pp 71-114.

Martin, R.W. (1981). Age-specific fertility in three populations of the koala, *Phascolarctos cinereus* (Goldfuss), in Victoria. Aust. Wildl. Res. 8, pp 275-83.

Martin, R.M. (1985). Overbrowsing, and decline of a population of the koala, *Phascolarctos cinereus*, in Victoria. I. Food preference and food tree defoliation. Aust. Wildl. Res. 12, pp 355-65.

Martin, R.M. (1985). Overbrowsing, and decline of a population of the koala, *Phascolarctos cinereus*, in Victoria. II. Population condition. Aust. Wildl. Res. 12, pp 367-75.

Martin, R.M. (1985). Overbrowsing, and decline of a population of the koala, *Phascolarctos cinereus*, in Victoria. III. Population dynamics. Aust. Wildl. Res. 12, pp 377-85.

McNally, J. (1957). A field survey of a koala population. Proc. Roy. Zool. Soc. N.S.W. 1955-56, pp 18-27.

McNally, J. (1960). Koala management in Victoria. Aust. Mus. Mag. 13, pp 178-81.

Minchin, A.U. (1937). Notes on the weaning of a young koala. Rec. S. Aust. Mus. 6, pp 1-3.

Oxlee, T.R. (1969). Victorian koala survey 1968. Wildl. in Aust. 6, p 56.

Parris, H.S. (1948). Koalas on the Lower Goulburn. Vict. Nat. 64, pp 192-3.

Philpott, C.M. (1965). The ecology of the koala, *Phascolarctos cinereus* (Goldfuss), on Flinders Chase, Kangaroo Island. B.Sc.(Hon.) thesis, University of Adelaide.

Robbins, M. & Russel, Eleanor (1978). Observations on movements and feeding activity of the koala in a semi-natural situation. in "The Koala" ed by T.J. Bergin. Taronga Park Zoo, Sydney, pp 29-41.

Smith, M.T.A. (1979). Notes on reproduction and growth in the koala, *Phascolarctos cinereus* (Goldfuss). Aust. Wildl. Res. 6, pp 5-12.

Smith, M.T.A. (1979). Behaviour of the koala, *Phascolarctos cinereus*, (Goldfuss), in captivity. I. Non-social behaviour. Aust. Wildl. Res. 6, pp 117-29.

Smith, M.T.A. (1979). Behaviour of the koala, *Phascolarctos cinereus* (Goldfuss), in captivity. II. Parental and infantile behaviour. Aust. Wildl. Res. 6, pp 131-140.

Smith, M.T.A. (1980). Behaviour of the koala, *Phascolarctos cinereus* (Goldfuss), in captivity. III. Vocalisations. Aust. Wildl. Res. 7, pp 13-34.

Smith, M.T.A. (1980). Behaviour of the koala, *Phascolarctos cinereus* (Goldfuss), in captivity. IV. Scent-Marking. Aust. Wildl. Res. 7, pp 35-40.

Smith, M.T.A. (1980). Behaviour of the koala, *Phascolarctos cinereus* (Goldfuss), in captivity. V. Sexual Behaviour. Aust. Wildl. Res. 7, pp 41-51.

Smith, M.T.A. (1980). Behaviour of the koala, *Phascolarctos cinereus* (Goldfuss), in captivity. VI. Aggression. Aust. Wildl. Res. 7, pp 177-90.

Warnecke, R.M., (1978). The status of the koala in Victoria. in "The Koala" ed by T.J. Bergin. Taronga Park Zoo, Sydney, pp 109-114.

Wicks, J.R. (1978). Koala preservation in an urban situation. in "The Koala" ed by T.J. Bergin. Taronga Park Zoo, Sydney, pp 148-151.

EVOLUTIONARY CONSIDERATIONS

Archer, M. (1976). Bluff Downs Local Fauna. Pp 383-95 in Archer, M. and Wade, M., (1976). Results of the Ray E. Lemley Expeditions, Part 1. The Allingham Formation and a new Pliocene vertebrate fauna from northern Queensland. Mem. Qd Mus. 17, pp 379-97.

Archer, M. (1976). Phascolarctid origins and the potential of the selenodont molar in the evolution of diprotodont marsupials. Mem. Qd Mus. 17, pp 367-71.

Archer, M. (1977). *Koobor notabilis* (De Vis), an unusual koala from the Pliocene Chinchilla Sand. Mem. Qd Mus. 18, pp 31-35.

Archer, M. (1978). Koalas (phascolarctids) and their significance in marsupial evolution. Pp. 20-8 in "The Koala" ed by T.J. Bergin. Taronga Park Zoo, Sydney.

Archer, M. (1981). A review of the origins and radiations of Australian mammals. Pp. 1437-88 in "Ecological biogeography of Australia" ed by A. Keast. Junk, The Hague.

Archer, M. (1982). Review of the dasyurid (Marsupialia) fossil record, phylogenetic interpretation, and suprageneric classification. In "Carnivorous marsupials" ed by M. Archer. Royal Zoological Society of New South Wales, Sydney.

Archer, M. and Hand, S.J. (1984). Background to the search for Australia's oldest mammals. In "Vertebrate zoogeography and evolution in Australasia" ed by M. Archer and G. Clayton. Hesperian Press, Perth.

Archer, M., Hand, S.J. and Godthelp, H. (1986). "Uncovering Australia's dreamtime". Surrey Beatty and Sons, Sydney.

Bartholomai, A. (1968). A new fossil koala from Queensland and a reassessment of the taxonomic position of the problematical species, *Koalemus ingens* De Vis. Mem. Qld Mus. 15, pp 65-73.

Bartholomai, A. (1977). The fossil vertebrate fauna from Pleisocene deposits at Cement Mills, Gore, southeastern Queensland. Mem. Qld Mus. 18, pp 41-51.

Bowler, J.M. (1982). Aridity in the late Tertiary and Quaternary of Australia. Pp. 35-45 in "Evolution of the flora and fauna of arid Australia, ed by W.R. Barker and P.J.M.

Greenslade. Peacock Publications, Frewville.

De Vis, C.W. (1889). On the Phalangistidae of the post-Tertiary period in Queensland. Proc. R. Soc. Qld 6, pp 105-14.

Hand, S.J. (1984). Bat beginnings and biogeography, a southern perspective. In "Vertebrate zoogeography and evolution in Australasia" ed by M. Archer and G. Clayton. Hesperian Press, Perth.

Hughes, R.L. (1965). Comparative morphology of spermatozoa from five marsupial families. Aust. J. Zool. 13, pp 533-43.

Jones, F. Wood. (1923). The monotremes and carnivorous marsupials. Part 1 in "The mammals of South Australia". Government Printer, Adelaide.

Kirsch, J.A.W. (1968). Prodromus of the comparative serology of Marsupialia. Nature, London 217, pp 418-20.

Kirsch, J.A.W. (1977). The comparative serology of Marsupialia, and a classification of marsupials. Aust. J. Zool. Suppl. Ser. 52.

Main, A.R. (1978). Ecophysiology, Towards an understanding of the late Pleistocene marsupial extinction. In "Symposium on biological problems in the reconstruction of Quaternary terrestrial environments" ed by D. Walker. Australian Academy of Science, Canberra.

Rich, T.H.V., Archer, M., Plane, M., Flannery, T.F., Pledge, N.S., Hand, S. and Rich, P.V. (1982). Australian Tertiary mammal localities. In "The fossil vertebrate record of Australasia" ed by P.V. Rich and E.M. Thompson. Monash Univ. Offset Printing Unit, Clayton.

Springer, M.S. (1987). The first lower molars of *Litokoala* (Marsupialia, Phascolarctidae) and their bearing on phascolarctid evolution. In "Possums and opossums, studies in evolution" ed by M. Archer. Surrey Beatty & Sons Pty Ltd and the Royal Zoological Society of New South Wales, Sydney.

Stirton, R.A. (1957). A new koala from the Pliocene Palankarinna Fauna of South Australia. Rec. S. Aust. Mus. 13, pp 71-81.

Stirton, R.A., Tedford, R.H. and Woodburne, M.O. (1967). A new Tertiary formation and fauna from the Tirari Desert, South Australia. Rec. S. Aust. Mus. 15, pp 427-62.

Strahan, R. (1978). What is a Koala? Pp. 3-19 in "The Koala" ed by T.J. Bergin. Taronga Park Zoo, Sydney.

Tedford, R.H. (1985). The Stirton years 1953-1966 a search for Tertiary mammals in Australia. Pp 38-57 in "Kadimakara" ed by P.V. Rich and G.F. van Tets. Pioneer Design Studio Pty Ltd, Lilydale.

Tyler, M. (1982). Tertiary frogs from South Australia. Alcheringa 6, pp 101-103.

Wells, R.T. and Callen, R.A. (eds) (1986). The Lake Eyre Basin—Cainozoic sediments, fossil vertebrates and plants, landforms, silcretes and climatic implications. Australasian Sedimentologists Group Field Guide Series No. 4. Geological Society of Australia, Sydney.

Woodburne, M.O., Tedford, R.H., Archer, M. and Pledge, N.S. (1987). *Madakoala*, a new genus and two species of Miocene koalas (Marsupialia, Phascolarctidae) from South Australia, and a new species of *Perikoala*. In "Possums and opossums, studies in evolution" ed by M. Archer. Surrey Beatty & Sons Pty Ltd and the Royal Zoological Society of New South Wales, Sydney.

THE KOALA AND MANKIND

Barrett, C. (1943) "Koala - The Story of Australia's Native Bear". Robertson & Mullens, Melbourne.

Fearn—Wannan, W. (1970). "Australian Folklore, A Dictionary of Lore, Legends and Popular Allusions". Lansdowne Press, Sydney.

Fountain, P. (1907). "Rambles of an Australian Naturalist". John Murray, Albermarle Street, London.

Gordon, G., and McGreevy, D.G. (1978). The status of the Koala in Queensland, pp 125-131 in T.J. Bergin (Ed). "The Koala", Proceedings of the Taronga Zoo Symposium. John Sands, Sydney.

Iredale, T., and Whitley, G. (1934). The Early History of the Koala. Victorian Naturalist 51.

Lewis, F. (1934). The Koala in Victoria. Victorian Naturalist 51.

Parris, H.S. (1948). Koalas on the Lower Goulburn. Victorian Naturalist 64, pp 192-93.

Peck, C.W. (1925). "Australian Legends". Stafford and Co Ltd., Sydney.

Portch, K. (1986). Ecology of the Koala in an Urban Environment. Unpublished B.Sc.(Hons) Thesis. University of New England.

Reed, A.W. (1965). "Myths and Legends of Australia". A.H. & A.W. Reed Pty Ltd. Sydney.

Reed, A.W. (1965). "Aboriginal Fables". A.H. & A.W. Reed Pty Ltd. Sydney.

Reed, A.W. (1978). "Aboriginal Legends". A.H. & A.W. Reed Pty Ltd. Sydney.

Ride, W.D.L. (1970). "Native Mammals of Australia". Oxford University Press. London.

Roberts, A. and Mountford, C.P.(1965). "The Dreamtime". Rigby Ltd. Adelaide.

Robinson, R. (1965). "The Man Who Sold His Dreaming". Currawong Publishing Co. Sydney.

Sutton, C.S. (1934). The Koala's Food Trees. Victorian Naturalist. 51, pp 78-80.

Troughton, E. (1941). "Furred Animals of Australia". Angus and Robertson, Sydney.

ACKNOWLEDGEMENTS

The line drawings, maps and reconstruction of the ancient rainforest environment of the Riversleigh area on page 95 were specially prepared for this edition by Sydney artist Mary Louise Brammer.
The work of the 19th century artist and naturalist John Gould appears on pp 11 and 12.

The photograph on the jacket and half-title page 1 are by L.F. Schick, that on page 64 is by H.J. Beste, both were supplied by the National Photographic Index of Australian Wildlife.
Photographs on pp 2, 63, 110 and 120 were kindly supplied by Qantas Airways Ltd.
Photographs on pp 42, 62, 74, 96 and 97 were kindly supplied by the NSW Dept of Tourism.

Other photographs were supplied by:
Michael Archer, pp 83, 84, 85, 86 and 98.
Steven Cork, p 29.
Stephen Phillips, pp 30, 39, 40, 41, 51, 61, 107, 108, 109 and 119.
Roger Martin, p 52.
Kath Handasyde, pp 29 and 73.

INDEX

Bold numerals indicate the main body of information on a subject.
Italic numerals refer to an illustration.